Editors

Prof. Dr. Gaston Berthier
Université de Paris
Institut de Biologie
Physico-Chimique
Fondation Edmond de Rothschild
13, rue Pierre et Marie Curie
F–75005 Paris

Prof. Dr. Michael J. S. Dewar
Department of Chemistry
The University of Texas
Austin, Texas 78712/USA

Prof. Dr. Hanns Fischer
Physikalisch-Chemisches Institut
der Universität Zürich
Rämistr. 76
CH–8001 Zürich

Prof. Kenichi Fukui
Kyoto University
Dept. of Hydrocarbon Chemistry
Kyoto/Japan

Prof. Dr. Hermann Hartmann
Akademie der Wissenschaften
und der Literatur zu Mainz
Geschwister-Scholl-Straße 2
D-6500 Mainz

Prof. Dr. Hans H. Jaffé
Department of Chemistry
University of Cincinnati
Cincinnati, Ohio 45221/USA

Prof. Joshua Jortner
Institute of Chemistry
Tel-Aviv University
61390 Ramat-Aviv
Tel-Aviv/Israel

Prof. Dr. Werner Kutzelnigg
Lehrstuhl für Theoretische Chemie
der Universität Bochum
Postfach 102148
D-4630 Bochum 1

Prof. Dr. Klaus Ruedenberg
Department of Chemistry
Iowa State University
Ames, Iowa 50010/USA

Prof. Dr. Eolo Scrocco
Via Garibaldi 88
I-00153 Roma

Prof. Dr. Werner Zeil
Direktor des Instituts
für Physikalische und
Theoretische Chemie
der Universität Tübingen
Aiblestraße 10
D-7406 Mössingen bei Tübingen

Lecture Notes in Chemistry

Edited by G. Berthier, M. J. S. Dewar, H. Fischer
K. Fukui, H. Hartmann, H. H. Jaffé, J. Jortner
W. Kutzelnigg, K. Ruedenberg, E. Scrocco, W. Zeil

14

Erik Waaben Thulstrup

Aspects of the
Linear and Magnetic Circular Dichroism
of Planar Organic Molecules

Springer-Verlag
Berlin Heidelberg New York 1980

Author

Erik Waaben Thulstrup
Department of Chemistry
Aarhus University
DK-8000 Aarhus C

ISBN-13: 978-3-540-09754-9   e-ISBN- 978-3-642-93136-9

DOI: 10.1007/ 978-3-642-93136-9

Library of Congress Cataloging in Publication Data
Thulstrup, Erik Waaben, 1941-
Aspects of the linear and magnetic circular dichroism of planar organic molecules.
(Lecture notes in chemistry; v. 14)
Bibliography: p.
Includes index.
1. Chemistry, Physical organic. 2. Circular dichroism. I. Title.
QD476.T45 548'.9 80-11592

2152/3140-543210

ASPECTS OF THE LINEAR AND MAGNETIC CIRCULAR DICHROISM OF
PLANAR ORGANIC MOLECULES

Erik W. Thulstrup
Department of Chemistry
Aarhus University
8000 Århus C, Denmark

CONTENTS

CONTENTS

I. INTRODUCTION: ASSIGNMENT CRITERIA

In the following the usefulness of the information that may be obtained from studies of the linear and magnetic circular dichroism of organic molecules will be demonstrated, and methods for increasing this information, especially from linear dichroism (LD), will be discussed. The LD of a sample is defined as the difference in absorbance found when linearly polarized light with the electric vectors in two directions, perpendicular to each other, is used, while the magnetic circular dichroism (MCD) is the difference in absorption of left and right circularly polarized light by a sample placed in a magnetic field.

Primarily the information obtained from LD and MCD tells something about the (electronic) transitions that cause the light absorption and about the corresponding excited electronic states. In order to maximize this information, the observed spectra must be related to results obtained from theoretical (quantum mechanical) models, although both LD and MCD have obvious applications as "primitive" analytical tools. It will be shown, however, that also the analytical applications may profit greatly from the use of simple as well as more elaborate quantum mechanical models for the description of LD and especially MCD. On the other hand, improvements in the theoretical models may appear as a result of the interaction with the information obtained from LD and MCD experiments.

The process relating the transitions observed in a spectroscopic experiment to the transitions calculated by means of a theoretical model is called an assignment (1,2). An assignment can be considered as a mapping M of a set of observed transitions $\{t_o\}$ into a set of calculated ones $\{t_c\}$. The mapping of an observed transition t_o is usually determined from a series of criteria C, each of which defines a subset of $\{t_c\}$, $\{t_c\}_C^o$, consisting of calculated transitions that according to C are possible mappings of t_o. If application of the available criteria, C_1, $C_2 \ldots C_N$, leads to a situation where the subsets $\{t_c\}_{C_i}^o$ have just one common element, this is the mapping of t_o:

$$Mt_o = \bigcap_{i=1}^{N} \{t_c\}_{C_i}^o$$

and t_o is assigned to Mt_o.

It should be noted that in practice one often prefers to treat only some of the observed transitions at a time (e.g. spin-allowed $\pi - \pi^*$ transitions in planar organic molecules) partly because of the nature of the theoretical models used. This requires a certain advance classification of the observed transitions which in most cases is quite straightforward.

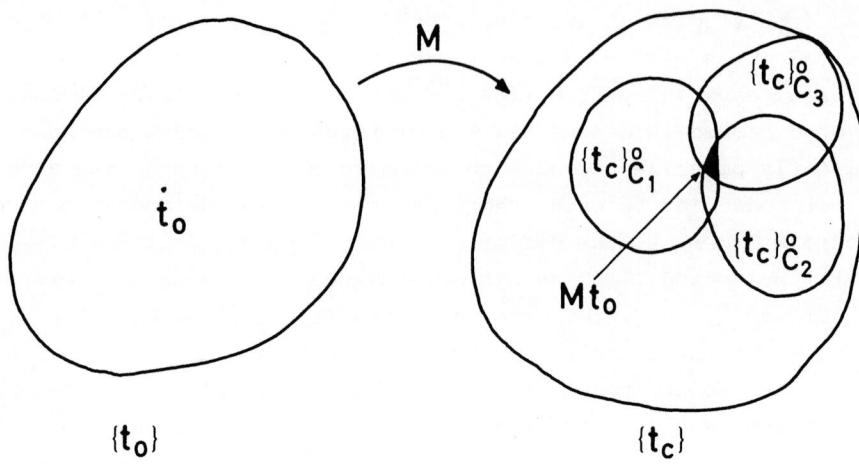

The most commonly used criteria, particularly in connection with electronic transitions in π-electron systems, are:

1. Transition energy.

For a typical semiempirical π-electron calculation this criterion might lead to:

$$\{t_c\}^o_{C_1} = \{t_c| \ |E(t_c)-E(t_o)|<7000 \ cm^{-1}\}.$$

2. Oscillator strength.

This criterion is often used qualitatively for separating allowed and forbidden transitions both with respect to spin and spatial symmetry. A typical more quantitative application of the criterion could lead to:

$$\{t_c\}^o_{C_2} = \{t_c|0.2<f(t_c)/f(t_o)<5\}$$

for $f(t_o) > 0.1$.

For weaker transitions wider limits must be used, both because of larger relative sensitivity of $f(t_c)$ to the method of calculation and because of the increased relative importance in the experimental spectrum of vibronic coupling, which is usually not included in the theoretical model, and which is difficult to evaluate accurately from the experiment.

3. Transition moment direction.

In the case of a molecule containing a symmetry plane the transition moment may be called a <u>strong</u> criterion. For instance, in molecules of C_{2v} or D_{2h} symmetry only three prependicular moment directions are possible. Thus, no estimate of the error limits is needed for the use of the criterion, and the subset of possible mappings $\{t_c\}^o_{C_3}$ of e.g. an observed transition t_o polarized along a direction $\phi(t_o)$ consists precisely of those transitions t_c which are calculated to be polarized in that direction. In cases where symmetry does not limit the number of possible transition moment directions, the criterion may lead to subsets of the form:

$$\{t_c\}^o_{C_3} = \{t_c| \ |\phi(t_c) - \phi(t_o)|<\pi/10\}$$

for $f(t_c) > 0.1$. $\phi(t)$ is supposed to be given as a single angle.

For weaker transitions the limit usually has to be wider for reasons similar to those given for the application of the oscillator strength criterion.

4. Magnetic circular dichroism.

Around 125 years ago Michael Faraday showed that the plane of polarization of linearly polarized light is rotated when the light passes through an absorbing sample placed in a magnetic field along the direction of the light beam.

This Faraday effect or MCD was not studied in much detail for the following 100 years, in spite of the fact that a similar effect, the natural optical activity (CD) was extremely popular and proved itself very useful in chemistry. In the 1960'ies the interest for MCD increased considerably. This interest was in particular directed towards molecules with degenerate electronic states for which the Zeeman splitting in a simple way accounts for the observed MCD effect. For example will the transitions from a non-degenerate ground state to the different components of a degenerate excited state give rise to the so-called A-term in the MCD spectrum. When the ground state is degenerate, an additional contribution appears, the temperature dependent C-term. Both these terms are simply related to the Zeeman splitting of degenerate states.

The vast majority of planar organic molecules does not belong to a point group with degenerate irreducible representations. Therefore, they do not have degenerate electronic states and their MCD spectra show no A- or C-terms. However, the MCD effect is found to be present in all molecules, and this is explained by the B-terms. These terms are not dependent on the splitting of degenerate states, but can be described as the result of a mixing of the electronic states by the magnetic field. This mixing appears both for the ground and excited electronic states as discussed in more detail later.

The magnetic circular dichroism of molecules without degenerate electronic states can thus be expressed by a single B-term for each transition. Since close-lying electronic transitions often have large B-terms of opposite signs, the criterion has been particularly useful for assignments of transitions in regions of two or three strongly overlapping bands (3,4). When the number of transitions becomes very large, as is often the case for π-electron systems in the short-wavelength UV region, MCD contributions from different transitions may be difficult to separate because of strong cancellation effects. Also, the theoretical results for such regions are often less reliable.

The B-terms of π-π^* transitions can for most molecules be calculated in the π-electron approximation with an accuracy that makes the MCD a very useful assignment criterion when the density of electronic states is reasonably small. In such regions with numerically large calculated and observed B-terms the application of a quantitative "B-term criterion" may lead to:

$$\{t_c\}^0_{C_4} = \{t_c | 0.2 < B(t_c)/B(t_o) < 5\}$$

which also ensures that the B-term sign criterion is fulfilled.

When two electronic states are calculated to be very close in energy, their B-terms may be strongly overestimated, and the upper limit should be increased. It may be added that the sign of a given calculated B-term only is meaningful when the term is of a reasonable size, which may be understood from the fact that the term often is calculated as a sum of positive and negative contributions.

5. Substituent effects.

The effect on the spectrum of the introduction of a substituent in the molecule or of a replacement (e.g. of C-H by nitrogen) may be a useful criterion, provided the change in the spectrum from the parent molecule to the derivative does not prevent recognition of the transitions. In other words: the substituent should act as a small perturbation, causing only minor changes in those parts of the ground and excited state wavefunctions that are important for a description of the transitions.

A well-known example of an application of the substituent effect criterion is provided by the benzene molecule, where the lowest two spin-allowed transitions (to $^1B_{2u}$ and $^1B_{1u}$) are spatially symmetry forbidden. A para-disubstitution will change the point group of the molecule from D_{6h} to (approximately) C_{2v} (D_{2h} for a symmetrical substitution) whereby $^1B_{2u}$ becomes $^1B_1$ and $^1B_{1u}$ becomes $^1A_1$. Transitions to these states are allowed and polarized perpendicular and parallel, respectively, to the axis between the substituents. Measurements of transition moment directions in para-$C_6H_4(OCH_3)_2$ (5-7), para-$C_6H_4(C_3H_7)_2$ (6), and para-$C_6H_4CH_3OH$ (6) indicate that the lowest singlet-singlet transition in benzene is to a state of $^1B_{2u}$-symmetry, while the second lowest is to a state of $^1B_{1u}$-symmetry.

Numerous other examples of the use of substituent effects exist, such as the investigation of the $^1B_{3u}^-$ transition (^1L_b) in anthracene (8) which in the unsubstituted alternant hydrocarbon is forbidden by the orbital pairing symmetry.

6. Less general criteria.

Other criteria may be very useful in special cases, but are not as generally applicable as the criteria discussed above. This group includes: vibrational structure, ESR splitting and ESR hyperfine structure of a long-lived upper state, solvent shifts, enhancement of singlet-triplet transitions by heavy-atom or paramagnetic substituents or solvents and many others.

II. EXPERIMENTAL TECHNIQUES

The experimental problems connected with LD measurements are generally much more complicated than those involved in ordinary MCD spectroscopy. The main reason is the need in LD studies for oriented molecular samples. MCD spectra can be obtained on commercially available instruments from liquid solutions in suitable solvents of the molecules to be investigated (9). Special applications of the MCD technique such as measurements on oriented samples are presently being attempted (10) and the theoretical basis for the orientational effects in such experiments has been worked out (11). An application with obvious analytical aspects (such as identification of compounds in very low concentrations) is fluorescence MCD, which has only recently been attempted experimentally, and for which some theoretical studies have been made (12).

Once the oriented sample has been obtained, experimental LD spectroscopy is relatively simple. It can be performed on an ordinary spectrophotometer equipped with a polarizer. An improved accuracy may be obtained by applying the phase modulation technique, which is used in circular dichroic (CD) spectroscopy (13). A discussion of this method has been given by Nordén et al. (14,15) and a general review of the theoretical background is contained in works by Jensen (16).

The methods currently used for obtaining oriented molecular samples are numerous. The present discussion will concentrate on the stretched sheet technique which may be the most efficient general method available for organic molecules with up to 50-100 atoms. Also other important methods will be briefly discussed. General reviews of these methods have been given by Dörr (17) and Nordén (15).

1. The stretched sheet method.

The use of anisotropic solvents for obtaining oriented samples is in principle a very straightforward method. Films and fibres have been used for many years in connection with the determination of transition moment directions both of elongated dye molecules, as well as of several other molecules (18,19). Among the numerous film materials, polyethylene (PE) seems to have some advantages, especially for non-polar molecules. For polar solutes the use of polyvinyl alcohol (PVA) sheets means better solubility, and PVA sheets have been used extensively. The solvent effects in polymer sheets are usually quite small.

A compound may be dissolved in e.g. PE sheets before or after stretching by immersing these in a suitable solution of the compound

(chloroform is a suitable solvent for PE), or by diffusion from the
vapor phase (20). The sheets are usually stretched 4-500% in one
direction. It is important that crystals on the surface of the film are
removed before the spectra are recorded; in some cases such crystals
have led to incorrect conclusions about the molecular spectrum (21).
If the solubility of a compound in the sheet is low, or if very weak
transitions are to be studied, a large number of sheets may be prepared
and pressed together before stretching at a temperature somewhat below
the melting point of the sheets (22). Also pieces of polyethylene
cut from stretched or extruded PE rods, and thereafter polished, are
very useful in these cases (23). It should be noted that it may take
days for the solute molecules to penetrate the thick sample to give
the maximum concentration.

In order to determine the absorption of the solute alone, spectra
of a pure stretched sheet must also be recorded, as described in detail
in (24), which contains a general description of the method. It seems
probable, however, that most of the apparent absorption of e.g. polyethylene sheets in the visible-UV region is really scattering, and that
a different measuring technique might remove the sheet baseline problem,
which generally is by far the largest source of error in the experiment.
Very recently attempts have been made to use photo-acoustic spectroscopy
(PAS) (25) in this connection. PAS has been known for a long time (26),
but only recently applications in chemistry have become more common (27).
The possible advantages of PAS in connection with stretched sheet
measurements are primarily that only absorbed light (and not scattered
light) is detected, and that PAS can be used for studies of samples
with optical densities over a wider range (down to 10^{-5} compared with
$\sim 10^{-3}$ as in standard methods). The latter property may significantly
reduce the concentration problems connected with stretched sheet
spectroscopy and may in principle make it possible to study very weak
transitions (singlet-triplet) and strong, symmetry allowed transitions
in the same sheet.

The experimental efforts in connection with the use of PAS in
studies of molecules in stretched sheets are reasonably moderate (28).
A piezo-electric transducer is fixed to the sheet and the absorption
of periodic, monochromatic, linearly polarized light (e.g. from a pulsed
laser) is recorded by means of the voltage created by the pressure and
temperature variations caused by the light absorption in the sheet. It
is crucial for the success of the experiment that the sample is kept
well sound insulated. Also the amount of light intensity is crucial for
the signal to noise ratio, and pulsed lasers seem to offer great advan-

tages in this connection.

The orientation effect obtained in stretched sheets is in general not perfect, as discussed later; it depends on the type of sheet and the compound being investigated. The orientation is also somewhat temperature-dependent; often a better orientation is obtained at low temperature (29). The main importance of lowering the temperature in stretched sheet spectroscopy is, however, the improved spectral resolution, and several methods have been used to attain this goal. The simplest way of ensuring a temperature of $77^{O}K$ is simply to immerse the stretched sheet into liquid nitrogen in a quartz Dewar and to record the spectrum through the nitrogen (29). If the construction of the sheet holder is so that bubbling occurs (insufficient heat insulation from outside the Dewar, rough surfaces) or if sufficiently pure liquid nitrogen is not available, good results have been obtained by placing the sample in the Dewar immediately above the nitrogen surface (30). More expensive methods are required if lower temperatures are needed, such as the use of a closed-cycle helium refrigerator (31). It may be added that a special advantage of low temperature spectroscopy is that evaporation of compounds with high vapor pressures at room temperature is reduced. Also strongly oxygen-sensitive compounds have been studied in stretched sheets placed in a Dewar with liquid nitrogen, without any problems (32).

2. Electric fields.

Molecules with non-zero dipole moments may be oriented in electric fields. Unfortunately, the orientation effect is usually very low; at best, the relative extinction differences are around 10^{-4}, and side effects such as changes in transition energies and moments due to the electric field may be of the same relative magnitude. Reviews of the technique have been given by Dörr (17) and Nordén (15) and pioneering work was carried out by Kuhn (33), Labhart (34), Liptay and others (35). Although the method has the advantage that the electric field may be varied (switched on and off), thereby making studies of the dynamics of the alignment possible, it is clear that the method is generally less useful for the determination of transition moment directions than e.g. the stretched sheet method.

3. Magnetic fields.

Very large systems (> 10^4 Å) may be oriented in a magnetic field (36) but the method is of little importance for organic molecules in general.

4. Hydrodynamic shear.

Non-spherical molecules in a solvent that moves rapidly in a specific direction relative to the walls of the container will show an LD effect. A review of such methods has been given by Wada (37). So far this method has been useful for macromolecules only and not for the determination of transition moment directions in ordinary organic molecules.

5. Crystal methods.

A single crystal may be considered the most natural case of oriented molecules. Unfortunately, the use of crystals in LD spectroscopy is often difficult. Besides the need for a complete knowledge of the crystal structure, the method requires very pure crystals and the investigation of intense transitions often makes extremely thin crystals necessary. There seems to be a possibility that the latter problem in some cases may be solved by means of the PAS technique mentioned above, but so far the mixed crystal technique has been the most common method used. This technique takes advantage of solute orientation in a host crystal consisting of molecules of similar shape, but transparent in the wavelength region of interest. Also in this case the crystal structure must be known, particularly the orientation of the guest molecules.

A different problem, especially in pure crystals, is the perturbation of the molecular states caused by the surroundings (38) which may change the absorption spectrum of the single molecule drastically, one reason why it is very important to remove crystals from the stretched sheet surfaces, as mentioned earlier. One significant advantage of the crystal methods should be mentioned: once the crystal structure is known, the transition moment directions may be found without further assumptions. In the stretched sheet method, for instance, it is usually assumed that the long axes in molecules tend to align with the stretching direction. This makes a comparison of results obtained by crystal techniques for single transitions in different molecules with those of other techniques very important, in spite of the fact that the general applicability of crystal methods for large organic molecules seems to be rather limited.

Early applications of the crystal methods were made by McClure (39), and reviews by Wolf (40) and Dörr (17) discuss the methods in more detail.

6. Liquid crystal methods.

Similarly to the stretched sheet or mixed crystal techniques the liquid crystal method makes use of the orientation of a solute in anisotropic solvents. Among the different liquid crystal systems used as solvents, uniformly aligned nematic phases are particularly important. The alignment of the liquid crystal may be obtained in several ways: by magnetic or electric fields, between quartz plates pressed together, by rubbing of the surfaces, by heating to the temperature range of the isotropic phase and then cooling slowly, or by treating the windows of the container with egglecithin. A review of these methods has been given by Sackmann (41) together with a description of spectroscopical applications of liquid crystals. A serious problem in connection with the method is the lack of suitable liquid crystal materials, transparent in the UV region. However, this problem may soon be reduced (42) and liquid crystals may become very important in UV LD spectroscopy.

7. Photoselection.

A different method for producing an oriented sample is to select a set of (partially) aligned molecules by means of excitation with linearly polarized light. In a glass, the molecules will preserve the alignment for some time and the polarization of the luminescence (fluorescence as well as phosphorescence) from the sample can be studied. The experiment may be performed using a fixed wavelength for the emitted light and varying the wavelength over part of or the whole range of the absorption (excitation spectrum), or vice versa (luminescence spectrum). Also, photochemical processes and other phenomena have been studied using this technique (17). By the photoselection method information is only obtained about the relative directions of the absorbing and emit-ting transitions, but it is nevertheless very useful, especially when combined with the results of other LD methods (43), as described later. Early applications of the photoselection method were made by Weigert (44) and important results were obtained by Zimmermann and Joop around 1960 (45). The basic formalism has been given by Albrecht and co-workers (46) as well as by Czekalla and Liptay (47).

The photoselection technique may also be applied to molecular samples already partially aligned (48,49,50). In particular, this application may provide new information about the orientation distribu-tion in stretched sheets, as discussed in the following chapter. The additional information (relative to that obtained from absorption experiments alone) is a result of the interaction of the molecule with two polarized photons (the absorbed and the emitted) instead of just one. For a uniaxial sample it is possible to record five independent spectra by varying the settings of the two polarizers relative to each other and to the (unique) sample axis.

It should be emphasized that the samples after they have been photoselected in general do not have a uniaxial alignment. Thus, application on the photoselected set of molecules of descriptions based on this symmetry property may lead to incorrect results.

III. INTERPRETATION OF THE OBSERVED SPECTRA

MCD and especially LD spectroscopy contains numerous examples of important experimental information that has been left unused or even severely misinterpreted (51). In this chapter a discussion will be given of some major problems involved in the understanding of observed LD and MCD spectra of organic molecules, and a general method for the interpretation of LD spectra will be described.

1. Vibronic interactions.

In the following it will be assumed that all transitions may be classified as electric dipole transitions and that magnetic dipole and electric quadrupole transitions are too weak to be observed in the absorption. Then the transition moment in the Born-Oppenheimer approximation for a transition between a state characterized by electronic and vibrational quantum numbers (i,v) to a state characterized by (j, u) may be written:

$$\vec{M}_{i,v}^{j,u} = <\Psi_i^{el\cdot}(q,Q)\,\chi_{i,v}^{vib\cdot}(Q)\,|\vec{M}(q,Q)\,|\,\Psi_j^{el\cdot}(q,Q)\,\chi_{j,u}^{vib\cdot}(Q)>_{q,Q}$$

where (q,Q) represent electronic and nuclear coordinates, respectively, and $\vec{M}(q,Q) = e\sum_p^{el\cdot}\vec{q}_p - e\sum_r^{nucl\cdot}z_r\vec{Q}_r$ is the dipole moment operator. The second sum vanishes in the expression for $\vec{M}_{i,v}^{j,u}$ due to the integration over the electronic coordinates. This integration defines the "electronic transition moment":

$$\vec{M}_i^j(Q) = <\Psi_i^{el\cdot}(q,Q)\,|\,e\sum_p^{el\cdot}\vec{q}_p\,|\,\Psi_j^{el\cdot}(q,Q)>_q$$

as discussed in many standard textbooks (52).

If $\vec{M}_i^j(Q)$ is nonzero for the equilibrium nuclear configuration Q_0 and is a slowly varying function of Q, the transition moment may be approximated by:

$$\vec{M}_{i,v}^{j,u} \simeq \vec{M}_i^j(Q_0)<\chi_{i,v}^{vib\cdot}(Q)\,|\,\chi_{j,u}^{vib\cdot}(Q)>_Q$$

The integral over nuclear coordinates expresses the Franck-Condon principle. For large organic molecules this usually means that unless the geometry of the molecule is drastically changed, a transition from the ground state corresponding to a totally symmetric (in the molecular point group symmetry) vibrational mode to an excited state (j,u) will only appear with large intensity if u also represents a totally symmetric

vibrational state. Therefore, the vibrational modes observed for allowed electronic transitions in absorption spectra can often be assumed to
correspond to totally symmetric normal coordinates.

The electronic transition moment is zero if the two states
involved have different spin or if the product of the irreducible
representations in the molecular point group of the electronic states
$\psi_i^{el.}$ and $\psi_j^{el.}$ does not contain the representation of one or more of
the translational coordinates. In this case the introduction of $\vec{M}_i^j(Q_0)$
is not satisfactory and "vibrational borrowing" (52) may become important. The theory for this effect has been given by Herzberg and Teller
(53) and is based on an expansion of the electronic wavefunction over
the electronic states $\{\psi_m^{el.}\}$:

$$\psi_j^{el.}(q,Q) - \psi_j^{el.}(q,Q_0) = \sum_{m\neq j}^{el.st.} c_m(Q)\psi_m^{el.}(q,Q_0)$$

$$c_m(Q) = \sum_{n}^{n.c.} [E_j^{el.}(Q_0) - E_m^{el.}(Q_0)]^{-1}<\psi_j^{el.}(q,Q_0)|(\frac{\partial H}{\partial Q_n})_{Q_0}Q_n|\psi_m^{el.}(q,Q_0)>_q$$

where the sum is over the normal coordinates $\{Q_n\}$. It is often assumed
that $\psi_i^{el.}(q,Q) = \psi_i^{el.}(q,Q_0)$ for the ground state. Then, for a forbidden
electronic transition ($\vec{M}_i^j(Q_0) = 0$) one has:

$$\vec{M}_{i,v}^{j,u} =$$

$$\sum_{m\neq j}^{el.st.} \sum_{n}^{n.c.} [E_j^{el.}(Q_0) - E_m^{el.}(Q_0)]^{-1}<\psi_j^{el.}(q,Q_0)|(\frac{\partial H}{\partial Q_n})_{Q_0}|\psi_m^{el.}(q,Q_0)>_q$$

$$\cdot <\psi_i^{el.}(q,Q_0)|e\sum^{el.}\vec{q}_p|\psi_m^{el.}(q,Q_0)>_q<\chi_{i,v}^{vib.}(Q)|Q_n|\chi_{j,u}^{vib.}(Q)>_Q . \qquad (III,1)$$

Thus a forbidden electronic transition to a state $\psi_j^{el.}$ may borrow
intensity from an allowed transition to a state $\psi_m^{el.}$ if the two
transitions are reasonably close in energy, and a suitable vibrational
normal mode exists which ensures that the first and third integral in
(III,1) are both nonvanishing. If $\vec{M}_i^j(Q_0)$ is zero due to point group
symmetry, a non-totally symmetric mode is necessary: if $\psi_m^{el.}(q,Q_0)$
forms a basis for the same irreducible representation as does
$\psi_j^{el.}(q,Q_0)$, the second integral in (III,1) will be zero, if not, the
first integral vanishes for all totally symmetric Q_n (6). Also in
cases of allowed, but weak electronic transitions effective vibrational
borrowing is observed between close-lying electronic states of different
symmetry (6). The borrowed intensity will, according to the expression
above, correspond to a transition moment direction along $\vec{M}_i^m(Q_0)$, and

will appear as vibronic peaks of the transition to $\psi_j^{el.}$ corresponding to the proper vibrational modes.

Usually the intensities and moment directions for electronic transitions in large molecules are calculated theoretically as the electronic transition moments $\{\vec{M}_i^j(Q_0)\}$ without inclusion of vibronic effects. If such are neglected, the observed extinction coefficient $\varepsilon_i^j(\nu)$ is related to $\vec{M}_i^j(Q_0)$ through the oscillator strength f_i^j:

$$f_i^j = 4.315 \cdot 10^{-9} \int \varepsilon_i^j(\nu)\,d\nu = 1.085 \cdot 10^{-5} \; \Delta E_{ij}^{el} \; \cdot \; \frac{|\vec{M}_i^j(Q_0)|^2}{e^2}$$

where ν is the wavenumber and the integration is carried out over the spectral region corresponding to the transition to $\psi_j^{el.}$. Other contributions to ε are supposed to have been removed.

2. Interpretation of LD-spectra. Absolute moment directions.

The observed linear dichroism provides information about the direction in the molecule of $\vec{M}_{i,v}^{j,u}$ and thus about the electronic transition moment $\vec{M}_i^j(Q_0)$. For molecules having two perpendicular symmetry planes only three possible directions for $\vec{M}_i^j(Q_0)$ exist. In such cases even crude experiments may give very precise information, in lower symmetry cases the information may be less valuable when large experimental errors are present.

It was mentioned above that the fluorescence polarization method is able to give information only about relative transition moment directions. Also other methods are limited in their ability to provide complete answers. In general, the stretched sheet and similar methods provide only values of the angles betweeen the transition moments and a fixed direction in the molecule, the so-called orientation axis (24), which will be discussed later. The possible transition moment directions are thus restricted to a cone around the orientation axis. Even when all transition moments are known to be in a plane ($\pi-\pi^*$ transitions in planar molecules) there are still in general two possible directions for each transition. In some cases, the unknown "sign" of the transition moment for transition j may be determined, e.g. from information on the angle α_j between the transition moment of j and the emitting transition obtained from a fluorescence polarization experiment. The angle α_j is given by (45,46):

$$\cos^2\alpha_j = (3P_j + 1)/(3 - P_j)$$

where $P = (I_Z - I_Y)/(I_Z + I_Y)$ is the degree of polarization; intensities observed when the polarizers used in the exciting and emitted beams are parallel and perpendicular are called I_Z and I_Y, respectively.

It was mentioned above that the stretched sheet method is able to provide the numerical value of the angle ϕ_j between the orientation axis and the transition moment of the jth transition. Assuming that the emitting transition is labelled 1 one has in a planar case (43):

$$\phi_j - \phi_1 = \pm \alpha_j \qquad\qquad j = 2, 3, 4, \ldots$$

where $|\phi_1|$, $|\phi_j|$ and $|\alpha_j|$ are known. For many transitions j only one set of signs for ϕ_j and α_j fits the equation, then the absolute moment directions are known, assuming that the position of the orientation axis and the sign for one transition has been established. In addition a more precise determination of $|\phi_1|$ is often obtained [43].

3. The description of a partially oriented uniaxial sample.

The molecular orientation in a stretched polymer sheet may be described by the three Euler angles (α, β, γ) (55) corresponding to the rotation of a molecule-fixed coordinate system (x, y, z) relative to that of a sheet-fixed coordinate system (X, Y, Z). In the former system the orientation axis z has been defined as the direction in the molecule along which a transition moment should be aligned in order to give the highest dichroic ratio E_Z/E_Y (24, 54, 55) where $E_Z(\lambda)$ and $E_Y(\lambda)$ are the observed absorbances with the stretching direction and electric vector of the light parallel and perpendicular, respectively. The direction perpendicular to z, which provides the highest possible dichroic ratio has been defined as y, and x was chosen perpendicular to z and y, ensuring a righthanded coordinate system. This choice of axes is identical to the one obtained by diagonalizing the 3 x 3 matrix $\{<\cos(iZ)\cdot\cos(jZ)>\}$ where the averaging is over the molecules to be studied, (i,j) are axes of a molecule-fixed coordinate system and Z is the stretching direction. The axis perpendicular to Z in the sheet plane is called Y, and X, perpendicular to Y and Z, is usually the direction of the light path in the experiment.

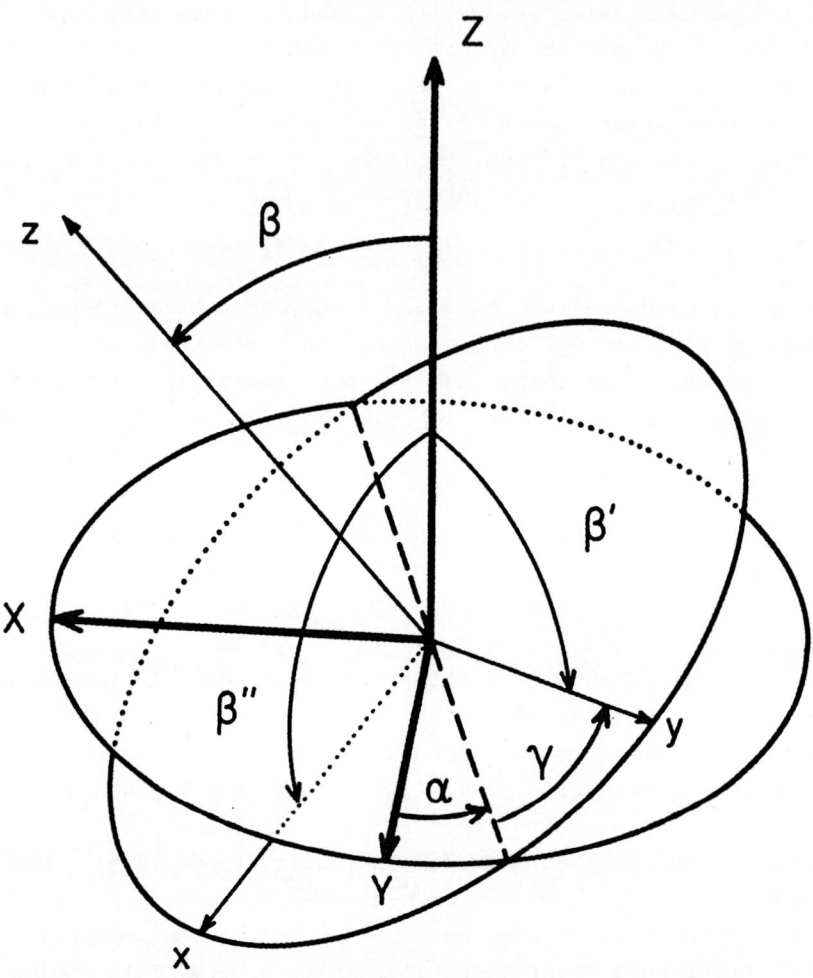

Sheet-fixed (X, Y, Z) and molecule-fixed (x, y, z) coordinate systems and the Euler angles (55).

For simplicity only coordinate system (x, y, z) will be used for the characterization of directional properties in the molecule. However, the use of a general coordinate system is straightforward (51, 56).

In the following it will be assumed that the distribution of the angle α is uniform. This corresponds to assuming rotational symmetry around the stretching direction, which simplifies the equations for interpretation of the spectra considerably. The assumption has been investigated experimentally for stretched polyethylene by X-ray diffraction (55) and seems to hold reasonably well. It is also assumed that the orientation is non-polar such that the values β = 0 and β = π are equally probable. The generalization to non-uniaxial or polar conditions is in principle simple, but requires additional parameters for the description of the various aspects of the orientation distribution. Expressions for the non-uniaxial case are given later.

The orientation distribution function $\rho(\alpha, \beta, \gamma)$ of solute molecules in a sheet may be expanded in a complete set of rotation matrices $D^j_{mm'}(\alpha, \beta, \gamma)$:

$$\rho(\alpha, \beta, \gamma) = \frac{1}{8\pi^2} \sum_{j,m,m'} (2j+1) <D^{j*}_{mm'}> D^j_{mm'}(\alpha, \beta, \gamma).$$

The rotation matrices are defined by:

$$D^j_{mm'}(\alpha, \beta, \gamma) =$$

$$e^{-i(m'\alpha+m\gamma)} \sum_\kappa \frac{(-1)^\kappa \sqrt{(j+m)!(j-m)!(j+m')!(j-m')!}}{\kappa!(j+m-\kappa)!(j-m'-\kappa)!(\kappa+m'-m)!} (-\tan\frac{\beta}{2})^{2\kappa+m'-m} (\cos\frac{\beta}{2})^{2j}$$

The coefficients $<D^{j*}_{mm'}>$ may in principle be determined from experiment; they contain the information about the orientation distribution. If the sample is perfectly aligned (all β = 0 or π) with uniform α and γ distributions, the coefficients become:

$$<D^{j*}_{00}> = 1 \text{ for j even, otherwise } <D^{j*}_{mm'}> = 0.$$

A different and very useful description of the orientation distribution is based on the angles β (= Euler angle β), β' and β" between the Z axis (stretching direction) and the molecular axes z, y and x, respectively (19, 55). The angles β' and β" are related to the Euler angles by the relations:

$$\cos\beta' = \sin\beta\sin\gamma, \quad \cos\beta" = \sin\beta\cos\gamma \tag{III,2}$$

in accordance with the fact that:

$$\cos^2\beta + \cos^2\beta' + \cos^2\beta'' = 1 \qquad\qquad (III,3)$$

In general an infinite number of coefficients $\langle D_{mm'}^j\rangle$ is necessary for a full description of the orientation distribution function. However, except in a few special cases, only a small number of coefficients can be determined experimentally. From an LD absorption experiment, (a one-photon process), information may be obtained about averages over the second power of directional cosines, e.g. $\langle\cos^2\beta\rangle$ and $\langle\cos^2\beta'\rangle$, corresponding to $\langle D_{00}^{2*}\rangle$ and $\langle D_{02}^{2*}\rangle + \langle D_{0-2}^{2*}\rangle$ (see below), fluorescence polarization, polarized Raman, and other two-photon processes may give information about the coefficients related to averages over the 4th power of the directional cosines, such as $\langle\cos^4\beta\rangle$ and $\langle\cos^2\beta\cos^2\beta'\rangle$, and a few additional coefficients can be obtained from e.g. NMR studies (57). A complete determination of $\rho(\alpha, \beta, \gamma)$ is at best feasible in special cases, e.g. when x-ray diffraction data can be obtained, which may be possible for solute molecules containing heavy atoms in suitable positions. The frequent claims that the orientation distribution has been determined on the basis of absorption (and maybe fluorescence) studies alone (49, 58) should not be taken seriously.

The averages over the cosine powers of the β's are related to the coefficients $\langle D_{mm'}^{j*}\rangle$. Some important relations are:

$$3\langle\cos^2\beta\rangle = (1 + 2\langle D_{00}^{2*}\rangle)$$

$$3\langle\cos^2\beta'\rangle = 1 - \langle D_{00}^2\rangle + (\sqrt{6}/2)(\langle D_{02}^{2*}\rangle + \langle D_{0-2}^{2*}\rangle)$$

$$5\langle\cos^4\beta\rangle = 1 + (20/7)\langle D_{00}^{2*}\rangle + (8/7)\langle D_{00}^{4*}\rangle$$

$$5\langle\cos^4\beta'\rangle = 1 + (5\sqrt{6}/7)(\langle D_{02}^{2*}\rangle + \langle D_{0-2}^{2*}\rangle) - (10/7)\langle D_{00}^{2*}\rangle +$$

$$+ (3/7)\langle D_{00}^{4*}\rangle - (\sqrt{10}/7)(\langle D_{02}^{4*}\rangle + \langle D_{0-2}^{4*}\rangle) +$$

$$+ (5/\sqrt{70}(\langle D_{04}^{4*}\rangle + \langle D_{0-4}^{4*}\rangle).$$

$$15\langle\cos^2\beta\cos^2\beta'\rangle = 1 + (5\sqrt{6}/14)(\langle D_{02}^{2*}\rangle + \langle D_{0-2}^{2*}\rangle) +$$

$$+ (5/7)\langle D_{00}^{2*}\rangle + (3\sqrt{10}/7)(\langle D_{02}^{4*}\rangle + \langle D_{0-2}^{4*}\rangle) -$$

$$- (12/7)\langle D_{00}^{4*}\rangle$$

In the important case of uniform distribution, the results are: $\langle\cos^2\beta\rangle = \langle\cos^2\beta'\rangle = 1/3$ and $\langle\cos^4\beta\rangle = \langle\cos^4\beta'\rangle = 1/5$. Various more recent descriptions of the orientation distribution in anisotropic solutions have been given by Gō (59), Kuball, Karstens, and Schönhofer (60), Linderberg (11a) and others (56). Linderberg points out that the information content of the distribution function $\rho(\alpha,\ \beta,\ \gamma)$ is defined as proportional to:

$$I = \ln(8\pi^2) + \int d\alpha\sin\beta d\beta d\gamma\rho(\alpha,\beta,\gamma)\,\ln\rho(\alpha,\beta,\gamma)$$

The constant has been determined by assuming that $I = 0$ for $\rho(\alpha,\beta,\gamma)$ = constant (no information).

If the information about $\rho(\alpha,\beta,\gamma)$ is contained in some average values, as described above, it seems reasonable to look for a distribution function that incorporates all the information available but no more. Therefore, I should be minimized under the constraint imposed by the experimentally determined average values such as $\{\langle D_{mm'}^{j*}\rangle\}$. This leads to the expression:

$$I = \overset{info.}{\underset{}{\Sigma}}\ a(jmm')\langle D_{mm'}^{j*}\rangle + \ln(8\pi^2)$$

where $a(jmm')$ are Lagrangian multipliers. Distribution functions with a larger amount of information cannot be considered acceptable even if they give the same set of average values. The main advantage connected with the use of the averages over the second and fourth powers of the directional cosines is that such averages correspond exactly to the information obtained in one- or two-photon experiments on oriented samples.

4. Determination of reduced spectra for symmetrical molecules.

Many important planar molecules belong to a symmetry point group (D_{2h}, C_{2v}) which limits the electronic transition moments to three perpendicular directions in the molecule. One of these will in such molecules necessarily be the effective orientation axis z.

For planar molecules the second in-plane direction for transition moments will usually be y, (corresponding to the second largest value for E_Z/E_Y) and the out-of-plane axis will be x. In some cases (approximately rod-shaped molecules) transitions along y and x may provide the same dichroic ratio; in such cases the in-plane axis is chosen as y.

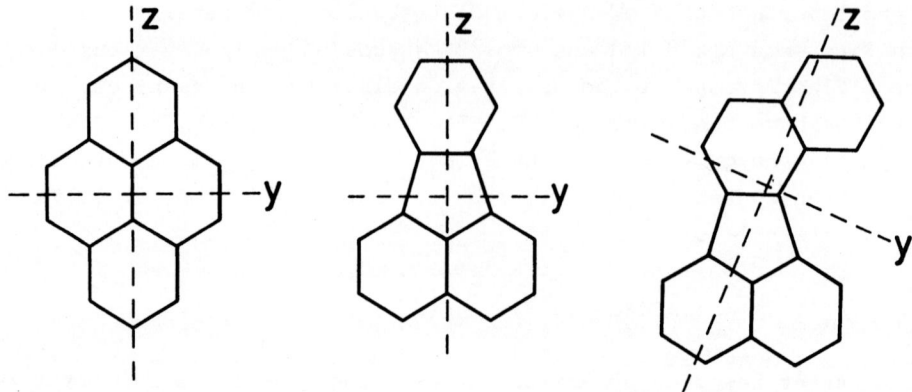

Planar aromatic hydrocarbons, corresponding to point groups D_{2h}, C_{2v} and C_s, respectively. The x-axes are perpendicular to the molecular planes. The electronic transition moments for the two first molecules must be along x, y or z, while they for the latter may be along x or along any direction in the (y,z)-plane.

The following notation has become standard in much of the LD literature:

$$\langle \cos^2 \beta \rangle = K_z$$

$$\langle \cos^2 \beta' \rangle = K_y$$

$$\langle \cos^2 \beta'' \rangle = K_x$$

From (III,3) one has:

$$K_x + K_y + K_z = 1$$

Thus the set of K's is linearly dependent. The observed absorbances can be written:

$$E_z(\lambda) = \sum_{i=x,y,x} K_i A_i(\lambda)$$

$$E_Y(\lambda) = \sum_{i=z,y,z} \frac{1}{2}(1-K_i) A_i(\lambda)$$

or

$$E_z(\lambda) = K_z A_z(\lambda) + K_y A_y(\lambda) + (1-K_z-K_y) A_x(\lambda) \qquad \text{(III,4)}$$

$$E_Y(\lambda) = \frac{1}{2}[(1-K_z)A_z(\lambda) + (1-K_y)A_y(\lambda) + (K_z+K_y)A_x(\lambda)]$$

Addition of these equations gives: $E_z(\lambda) + 2E_Y(\lambda) = A_z(\lambda) + A_y(\lambda)$ + $A_x(\lambda) = 3A(\lambda)$, where $A_z(\lambda)$, $A_y(\lambda)$ and $A_x(\lambda)$ are absorption components polarized along the three axes and $A(\lambda)$ the absorption corresponding to an isotropic sample. This result is expected, since the uniaxial property means that $E_x(\lambda) = E_Y(\lambda)$. In some cases other quantities than $E_z(\lambda)$ and $E_Y(\lambda)$ are measured, such as $E_z(\lambda) + E_Y(\lambda)$ and $E_z(\lambda) - E_Y(\lambda)$. This does not change the following expressions, but in some cases simple reformulation may be useful.

In order to obtain information about the curves $A_x(\lambda)$, $A_y(\lambda)$, and $A_z(\lambda)$, linear combinations of $E_z(\lambda)$ and $E_Y(\lambda)$ may be constructed so that spectral features due to z, y or x polarized transitions disappear (e.g. $\Delta(E_z(\lambda) - d_z E_Y(\lambda))/\Delta\lambda = 0$ over a spectral region for which contributions from y- and x-polarized transitions are constant or negligible). It has been emphasized (6,19,61) since the first applications of this method, that the requirement of pure polarization along one of the axes is not necessary, only spectral features (peaks, shoulders, or parts thereof) must be assumed purely polarized. Unfortunately, this point is still frequently misunderstood.

If such purely polarized spectral features can be found, it is possible to determine the linear combinations $E_z - d_z E_y$, $E_z - d_y E_Y$ or $E_z - d_x E_Y$ (the wavelength dependence is in the following only explicitly indicated when necessary) where contributions from z, y, or x-polarized transitions, respectively, have been removed. Then from (III,4):

$$\langle\cos^2\beta\rangle = K_z = d_z/(2+d_z),$$

$$\langle\cos^2\beta'\rangle = K_y = d_y/(2+d_y),$$

$$\langle\cos^2\beta''\rangle = 1-K_z-K_y = K_x = d_x/(2+d_x).$$

Thus, K_i can be determined from the knowledge of d_i. Furthermore, these relations have been used to obtain an independent check of the results for a molecule with observed transitions polarized in three perpendicular directions (62). They also permit a determination of e.g. d_x when d_z and d_y are known.

From the two observed absorption curves, E_z and E_y, the components A_z, A_y and A_x cannot in general be determined. However, in some special cases very precise information may be obtained. In the LD literature this possibility has frequently been overlooked and often only part of the information available has been extracted.

One of the most important cases is that of $A_x(\lambda) = 0$ (e.g. π-π*-transitions in planar molecules). Albrecht attempted several years ago to estimate curves corresponding to transition moments in two perpendicular directions, z and y, from fluorescence polarization experiments (63), where the orientation distribution of the excited molecules is known. Later it was shown [19] how A_z and A_y may be obtained from the quantities E_z and E_y observed in stretched sheets, where the distribution function is unknown. Thus d_z and d_y must be determined experimentally from spectral features polarized along these two directions as described above; if such features are not available, assumptions about the size of K_z and K_y must be made. From the knowledge of (K_z, K_y) or (d_z, d_y) and from (III,4) "reduced" spectra are obtained:

$A_x(\lambda) = 0$
_____

$$A_z(\lambda) = [E_z(\lambda) - d_y E_y(\lambda)](2 + d_z)/(d_z - d_y) =$$

$$= [(1 - K_y)E_z(\lambda) - 2K_y E_y(\lambda)]/(K_z - K_y)$$

$$A_y(\lambda) = [d_z E_y(\lambda) - E_z(\lambda)](2 + d_y)/(d_z - d_y) =$$

$$= [2K_z E_y(\lambda) - (1-K_z)E_z(\lambda)]/(K_z - K_y) \tag{III,5}$$

An example is shown in Fig. 1.

Another special case is that of a "disc-shaped" molecule. Then, $K_z = K_y (d_y = d_z; \ d_x = (2-d_z)/d_z)$, and (III,4) leads to the following relation (54):

$d_y = d_z$
_____

$$A_x(\lambda) = [2E_y(\lambda)-(1+d_x)\,E_z(\lambda)]/(1-d_x) = [E_z(\lambda)-d_z E_y(\lambda)]/(1-d_z) =$$

$$= [2K_zE_Y(\lambda) - (1-K_z)E_Z(\lambda)]/(3K_z-1)$$

$$A_Y(\lambda)+A_Z(\lambda) = 2[E_Z(\lambda)-d_xE_Y(\lambda)]/(1-d_x) = [(2-d_z)E_Y(\lambda)-d_zE_Z(\lambda)]/(1-d_z) =$$

$$= 2[K_zE_Z(\lambda) - (1-2K_z)E_Y(\lambda)]/(3K_z-1) \tag{III,6}$$

This means that both $A_x(\lambda)$ and $A_y(\lambda)+A_z(\lambda)$ may be determined from the knowledge of only one of the quantities d_x or $d_z(=d_y)$.

A very important case is that of a "rod-shaped" molecule $(K_y = K_x = (1-K_z)/2; d_x = d_y = 2/(1+d_z))$. Substitution into (III,4) gives (54):

$$\underline{d_y = d_x}$$

$$A_x(\lambda)+A_y(\lambda) = [(2-d_x)E_Y(\lambda)-d_xE_Z(\lambda)]/(1-d_x) = 2[d_zE_Y(\lambda)-E_Z(\lambda)]/(d_z-1) =$$

$$= 2[2K_zE_Y(\lambda) - (1-K_z)E_Z(\lambda)]/(3K_z-1)$$

$$A_z(\lambda) = [E_Z(\lambda)-d_xE_Y(\lambda)]/(1-d_x) = [(1+d_z)E_Z(\lambda)-2E_Y(\lambda)]/(d_z-1) =$$

$$= [(1+K_z)E_Z(\lambda)-2(1-K_z)E_Y(\lambda)]/(3K_z-1) \tag{III,7}$$

The assumption of a rod-like orientation distribution thus makes it possible to determine A_x+A_y and A_z (or in the case of $A_x \sim 0$, A_y and A_z) form the knowledge of only one of the constants d_z or $d_y(=d_x)$.

In a general case A_x, A_y or A_z cannot be determined from only two measurements. At least three linearly independent results are necessary. Since $E_Z + 2E_Y = A_x + A_y + A_z = 3A$, a randomly oriented sample will not provide the curve needed in addition to E_Z and E_Y. The LD spectrum of a differently oriented sample (E_Z' and E_Y') may do, provided that:

$$\begin{vmatrix} K_z & K_y & K_x \\ K_z' & K_y' & K_x' \\ 1/3 & 1/3 & 1/3 \end{vmatrix} \neq 0$$

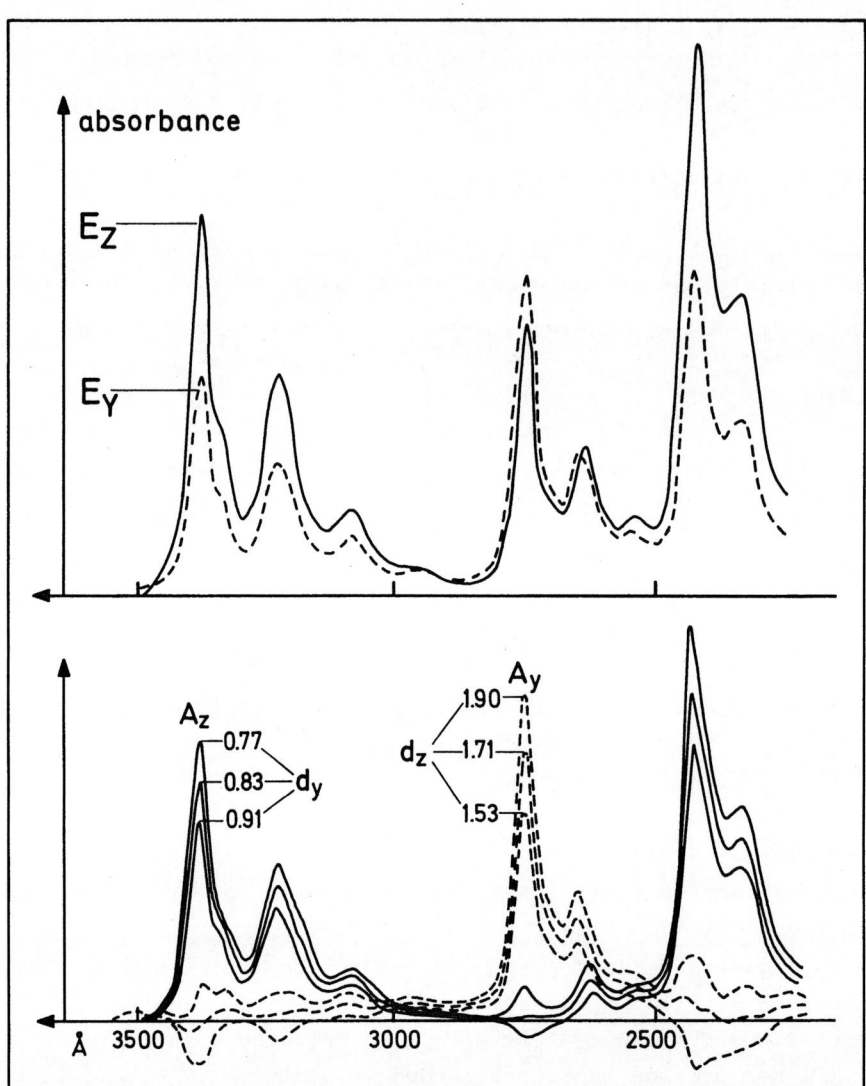

Fig. 1. Pyrene (19). Top: Observed spectra obtained at room temperature in stretched polyethylene sheets. Bottom: Linear combinations of E_Z and E_Y used for a determination of (d_z, d_y). The set (1.71, 0.83) should be chosen. The negative value around 3400 Å for the A_y-curve corresponding to $d_z = 1.71$ is assumed to be a result of a different "solvent shift" in the experimental curves E_Z and E_Y. Similar effects have been observed for many other molecules, see e.g. (58).

Table 1. Relations between orientation constants.

Orientation constants	K_i	d_i
$K_i = \langle \cos^2(Zi) \rangle$ $i = x, y, z$	$K_z \geq K_y \geq K_x \geq 0;$ $K_x + K_y + K_z = 1$	$K_i = d_i/(2+d_i)$
d_i; determined from e.g. $\Delta(E_z - d_i E_y)/\Delta\lambda = 0$ for an i-polarized spectral feature	$d_i = 2K_i/(1-K_i)$	$d_z \geq d_y \geq d_x \geq 0;$ $d_x/(2+d_x) + d_y/(2+d_y) + d_z/(2+d_z) = 1$
Disc-like case	$K_z = K_y \geq K_x$ $f^{a)} = 2(3K_z - 1)$	$d_z = d_y \geq d_x$
Rod-like case	$K_x = K_y \leq K_z$ $f^{a)} = (3K_z - 1)/2$	$d_x = d_y \leq d_z$
Perfect plane alignment	$K_x = 0; K_z + K_y = 1$	$d_x = 0; d_y d_z = 4$

a) f refers to applications of the Fraser-Beer model. See section 6a.

5. The orientation triangle.

A convenient way of illustrating the orientation of molecules in aniosotropic solvents and the relation between the K's is the orientation triangle (Fig. 2). The sides of the triangle, which describes the limiting values of (K_z, K_y) are:

$K_y = K_z, K_x = (1-K_z)/2$ (disc-like orientation, $d_z = d_y$)
$K_y = (1-K_z)/2 = K_x,$ (rod-like orientation, $d_y = d_x$)
$K_y = 1-K_z, K_x = 0,$ (perfect alignment of the molecular planes with the stretching direction, $d_x = 0$).

The lower left vertex (1/3, 1/3) may be obtained from a randomly orientated sample, the upper vertex (1/2, 1/2) from a sample of perfectly aligned disc-shaped molecules, while the lower right vertex (1,0) is obtained from perfect alignment of the orientation axis. It is of interest to note that the point (1,0) is the only point in the triangle that corresponds to only one orientation distribution (perfect alignment of the z-axis and the uniaxial property leave no degrees of freedom). All other points correspond to an infinite number of different orientation distributions.

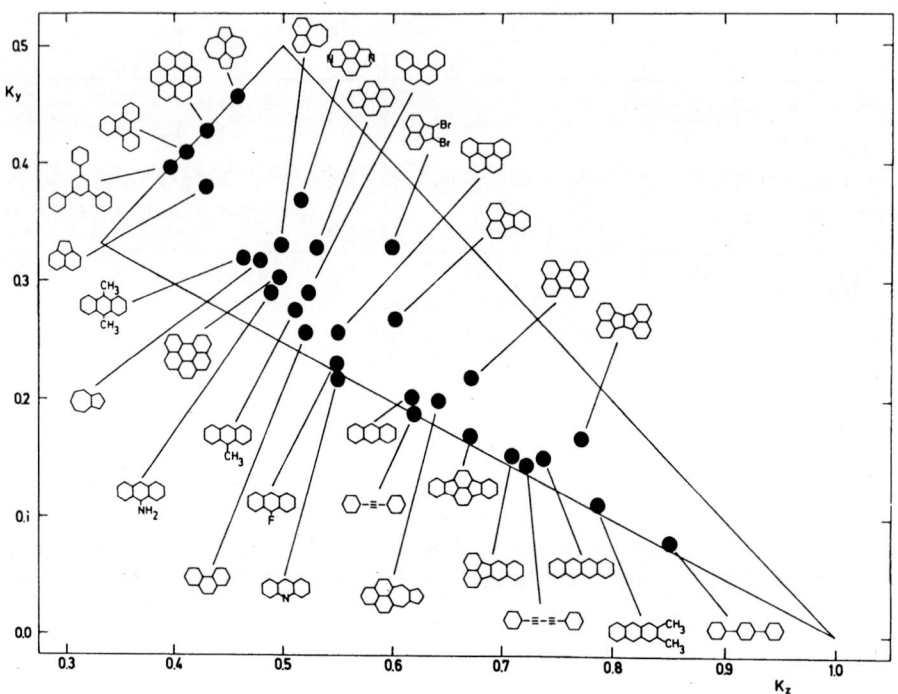

Fig. 2. The orientation triangle (54) including a number of experimental
(K_z, K_y) values. The lower side of the triangle describes rod-like
orientation distributions $(K_y = K_x)$, the upper left side disc-like
orientation $(K_z = K_y)$, and the upper right side corresponds to perfect
alignment of the molecular plane with the stretching direction
$(K_z + K_y = 1, K_x = 0)$. The K_x-value for any point (K_z, K_y) may be found
as the vertical distance between the point and the upper right side
of the triangle.

The orientation triangle was introduced several years ago in a less convenient form (7), and has been improved more recently (54,64). A different triangle has been used in the polymer literature for a description of the alignment of the polymer molecules (65). For the description of experiments, for which higher orders of the directional cosines are needed(polarized Raman, fluorescence polarization etc.), corresponding figures in higher dimensional spaces may be used (56).

A useful description similar to the one based on the average values of the cosine squares of the directional angles (the K's) has been introduced by Saupe for NMR-work on liquid crystals (66). Saupe's S-values are related to the K's by:

$$S_{ii} = (3K_i - 1)/2 \qquad i = x,y,z$$

$$\text{(III,8)}$$

$$K_i = (2S_{ii} + 1)/3 \qquad i = x,y,z$$

where $S_{xx} + S_{yy} + S_{zz} = 0$. This description also assumes a uniaxial orientation distribution (uniform distribution of the Euler angle α). An orientation triangle similar to the one defined for the K's may also be introduced for e.g. S_{zz} and S_{yy}.

6. Orientation models.

Unfortunately, a large number of papers on the orientation of molecules in anisotropic solvents have been based on incompletely defined assumptions, which frequently are incorrect or at best doubtful. It is usually assumed that the orientation distribution has some specific form which may be correct for some samples, but not in the general case (51). In view of the increasing interest in LD spectroscopy, this situation is very unfortunate and in the following some of the most common orientation models will be critically analyzed. The purpose of this investigation is exclusively to help remove some of the confusion that is found in the LD literature. It must be emphasized that many results which have been obtained by means of incorrect orientation descriptions are based on valuable experimental information and that the experimental contributions often have been of great importance for the development of the field.

6a. Assumption of a rod-like distribution. The Fraser-Beer Model.

In their work on the orientation of polymer chains Fraser and Beer (67,68) assumed that all angles of rotation of the chain around itself were equally probable. Their formalism has later been used for

work on solute molecules. For these, the assumption seems reasonable
only in the case of rod-shaped molecules. The model may correspond to
assuming that a fraction f of the dissolved molecules is perfectly
aligned (corresponding to $(K_z, K_y) = (1,0)$) while the rest $(1-f)$ is
oriented perfectly random $((K_z, K_y) = (1/3, 1/3))$. Then by varying f
from 0 to 1 all points on the lower side of the triangle in Fig. 2
can be reached. However, except for $(1,0)$, each of these points correspond
to an infinity of orientation distribution functions, one of which
corresponds to the f and $(1-f)$ distribution. It is obvious that the
model permits a description of any one-photon process in rod-shaped
molecules, but it is likely that a two-photon process such as fluorescence polarization, yielding information on averages of the fourth
powers of the directional cosines might not be described by the model,
unless the actual orientation distribution happens to correspond to
that assumed in the model. Fig. 3 shows the relative theoretical limits
for $\langle\cos^2\beta\rangle$ and $\langle\cos^4\beta\rangle$ as well as the curve corresponding to the
Fraser-Beer model. In this model with the above mentioned assumptions:

$$\langle\cos^2\beta\rangle = f \cdot 1 + (1-f) \cdot 1/3 = (1+2f)/3$$

$$\langle\cos^4\beta\rangle = f \cdot 1 + (1-f) \cdot 1/5 = (1+4f)/5$$

where $\langle\cos^4\beta\rangle$ and $\langle\cos^2\beta\rangle$ for a random distribution are 1/5 and 1/3,
respectively. The result is after elimination of f:

$$\langle\cos^4\beta\rangle = (6\langle\cos^2\beta\rangle - 1)/5 \qquad\qquad\qquad (III,9)$$

Thus points corresponding to a Fraser-Beer orientation distribution
lie on a straight line in the $\langle\cos^2\beta\rangle$, $\langle\cos^4\beta\rangle$ diagram.

In the figure are also shown the experimental results for
$(\langle\cos^2\beta\rangle, \langle\cos^4\beta\rangle)$ for 4,4'-(dibenzo-xazolyl)-stilbene, obviously a
rod-shaped molecule, in stretched poly(ethylene terephthalate) (69).
It is clear that the experimental points do not agree with the theoretical
line.

A problem in connection with the application of the Fraser-
Beer model which so far has been considerably more serious, is the
frequent use for descriptions of molecular orientations that cannot be
expected to be rod-like. This has often been done without even mention-
ing the crucial assumption, and the approximation has frequently led to
obvious errors, as well as results that may be considered very doubtful.
A single example may be mentioned: A study of fluoranthene in stretched

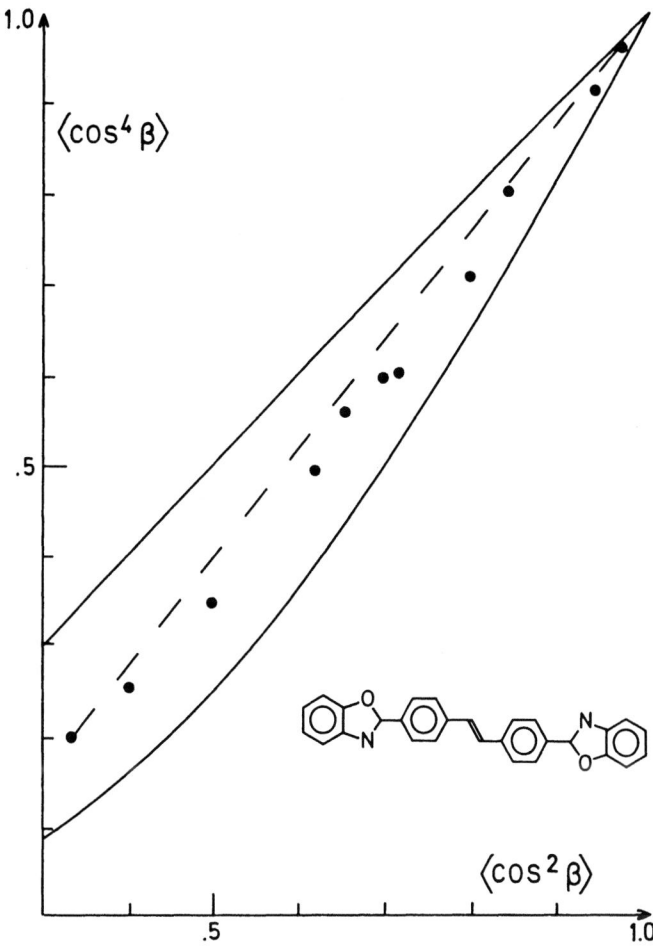

Fig. 3. The relation between $<\cos^4\beta>$ and $<\cos^2\beta>$. The two full lines indicate the theoretical upper ($<\cos^4\beta> = <\cos^2\beta>$) and lower ($<\cos^4\beta> = <\cos^2\beta>^2$) limits, and the dotted line corresponds to a Fraser-Beer (III,9) orientation ($<\cos^4\beta> = (6<\cos^2\beta>-1)/5$). The points indicate observed values obtained in stretched poly(ethylene-terephtha-late) (69) for the rod-shaped molecule 4,4'-(dibenzoxazolyl)stilbene shown in the figure.

polyethylene (70) resulted in reduced spectra with mixed polarization
of the y-polarized component, in particular the strong transition near
240 nm. Such mixed polarization is obviously very unlikely for the C_{2v}
molecule. It was emphasized (70) that the results were obtained from
only one assumption about the K's; however, the assumption of a rod-
like orientation distribution for fluoranthene was made implicitly.
The actual orientation distribution in the experiment did not have this
property (Fig. 2) and application of the Fraser-Beer model led to in-
correctly reduced spectra. Other examples of applications of the Fraser-
Beer model, where the rod-like orientation distributions assumed seem
highly unlikely, are found in studies of carotenoids(71,72), and
several other molecules (73-75). Examples of clearly incorrect assumptions are found for biliverdin (76) and (6,6)-vespirine (77). In the
latter case an "overreduction" seems to be present in the short axis
polarized component $A_y(\lambda)$, which contains a mirror image of the strong
z-polarized peak near 35000 cm^{-1}. Examples of questionable applications
of the Fraser-Beer model also occur in studies in the IR region (75).
It must be added that sometimes the assumption of a rod-like
orientation seems to be an acceptable approximation, e.g. when $(1-K_z)/2$
$\leq K_y \leq (1-K_z)/2 + 0.03$, and the results are likely to be approximately
correct (77-80).

It is interesting to note that in some cases (e.g. for acenapht-
ylene (64), see Fig. 2) y-polarized transitions are found to have a
dichroic ratio E_Z/E_Y larger than one. This corresponds to $K_z \geq K_y \geq 1/3$.
The Fraser-Beer assumption of $K_x = K_y$ then leads to $K_x + K_y + K_z > 1$,
which directly proves that the model does not work.

The Fraser-Beer-model has been defended by some authors (49, 58,
71, 81, 82) by suggesting that since the polyethylene frequently used
in LD experiments is composed of two regions, a crystalline and an
amorphous, it is probable that the molecules incorporated are either
perfectly oriented in the crystalline parts of the polyethylene or
randomly oriented in the amorphous part. This assumption does not seem
reasonable, since the f-values used often are large (>0.5) and such
high concentrations of guest molecules in the crystalline regions are
very unlikely (83). Moreover, since f is usually much larger for
elongated than for short molecules, the assumption imply that the
former molecules tend to go into the crystalline regions much more
easily than the latter, which seems highly improbable.

It is important to keep in mind that the experimental results
available on the alignment of molecules in stretched polymers (absorption, polarized Raman, fluorescence, magnetic resonance, etc.) do not
allow a complete determination of the orientation distribution and that

such a determination is not necessary for the interpretation of for example LD spectra. For one-photon processes like absorption, only information corresponding to the K's can be found and these properties of the orientation distribution are sufficent for the evaluation of the spectra. As mentioned before, one should look for a description of the orientation that accounts for all observed results, but does not involve additional assumptions.

The Fraser-Beer model has also been used by Nordén (84) and others (85) for disc-shaped molecules by assuming that such molecules tend to be aligned with their planes parallel to the stretching direction. While the Fraser-Beer model in a rod-like case leads to:

$$E_Z(\lambda) = [f \cos^2\alpha(\lambda) + (1-f)/3]A(\lambda)$$

$$\text{(III,10)}$$

$$E_Y(\lambda) = [(f/2) \cdot \sin^2\alpha(\lambda) + (1-f)/3]A(\lambda)$$

where $\alpha(\lambda)$ is the angle between the orientation axis and the transition moment direction at wavelength λ, the expression for a disc-shaped molecule becomes (84):

$$E_Z(\lambda) = [(f/2)\cos^2\alpha(\lambda) + (1-f)/3]A(\lambda)$$

$$\text{(III,11)}$$

$$E_Y(\lambda) = [(f/2)\sin^2\alpha(\lambda) + (f/4)\cos^2\alpha(\lambda) + (1-f)/3]A(\lambda)$$

Here $\alpha(\lambda)$ is the angle between the molecular plane and the transition moment. The parameter f can be considered being the fraction of the molecules which have their planes perfectly aligned with the stretching direction, while the fraction 1-f consists of randomly oriented mole- cules. Variation of f between 0 and 1 produces points on the left side of the orientation triangle in Fig. 2.

The use of (III,11) is limited to cases of perfect or almost perfect disc-like orientation distributions. When applied to samples without this property, incorrect conclusions will inevitably be reached. This may be the reason for the disagreement between the values found for the transition moment directions in hexahelicene by the stretched sheet (85) and by crystal methods (86).

The physical meaning and limitations of the Fraser-Beer model is well illustrated by the orientation triangle (Fig. 2). It is obvious that only few experimental points lie on the lower side of the triangle which corresponds to rod-like orientations, and only disc-shaped

molecules are represented on the left side of the triangle, which
corresponds to disc-like orientation distributions. A further illustra-
tion is obtained by noting that e.g. the assumption of rod-like distribution functions corresponds to assuming equivalency between different
values of the Euler angle γ:

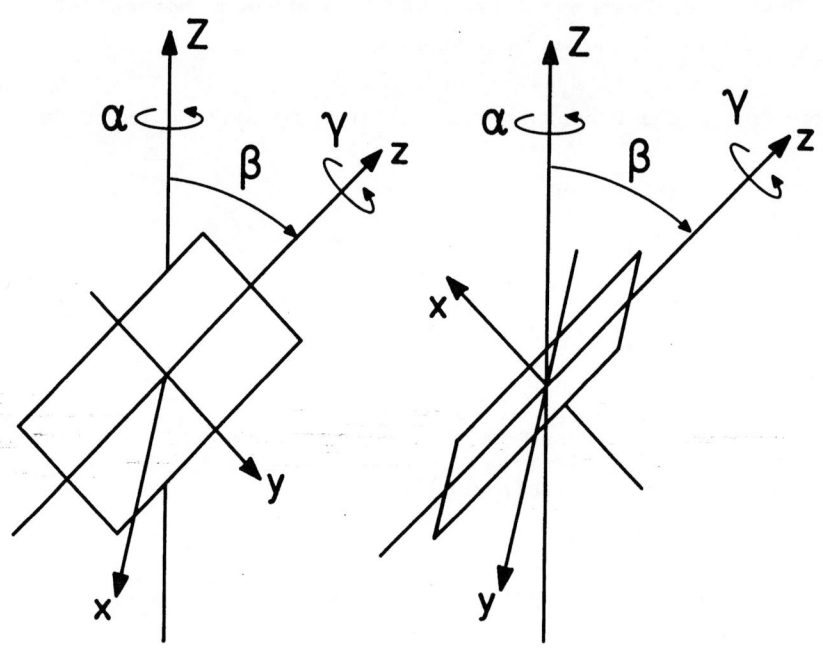

The two molecular orientations are obviously physically different,
except when the molecular axes y and x are equivalent.

A recent attempt to extend the Fraser-Beer model to molecules
which cannot be assumed to orient in a rod-like (or disc-like) way has
been made by Yogev and coworkers (49, 58, 82). The extended model
suggests that planar guest molecules in a polymer may be divided into
a fraction (f) of molecules, which have their planes parallel to the
stretching direction (upper side of the triangle), with a Boltzmann-
like distribution of the orientation axes, and a fraction (1-f) of
randomly oriented molecules. Special cases are the rod-shaped and disc-
shaped molecules. Again, this is an attempt to specify the orientation
distribution function in more detail than needed for the evaluation of

the spectra. It is obvious, according to the earlier discussion, that polarized absorption, Raman and fluorescence experiments cannot provide a proof for such a specific orientation distribution function; at best counter-proofs may be avoided. Although the proposed model (82) contains more flexibility and therefore may cause fewer obvious problems than the simple Fraser-Beer model, it seems equally unlikely from a physical point of view. Moreover, the model requires more elaborate assumptions than the general description proposed by Eggers, Michl and Thulstrup (6, 19, 24, 51), it is more complicated to use and offers no advantage over the latter model, as long as the assumed orientation distribution function cannot be shown to be close to the real one.

6b. Assumption of perfect plane alignment (Tanizaki's model).

One of the first orientation models was proposed by a leading pioneer in stretched sheet spectroscopy, Y. Tanizaki, and his coworkers (87-90). His model is for planar molecules based on the assumption of perfect alignment of the molecular plane with the stretching direction in a hypothetical infinitely stretched polymer (polyvinylalcohol, PVA, is the commonly used material). The assumed relation between the dichroic ratio that would be obtained at infinite stretching, $R_{d\infty}$, and the measured ratio $R_d = E_Z/E_Y$ is (91):

$$R_{d\infty}(\lambda) = \frac{2(T-1) + (T+1)R_d(\lambda)}{2T + (T-1)R_d(\lambda)} \tag{III,12}$$

where

$$T = \frac{R_s^2}{R_s^2-1} [1-(\pi/2-\tan^{-1}(R_s^2-1)^{-1/2})(R_s^2-1)^{-1/2}] \tag{III,13}$$

Here R_s is the stretch ratio, defined as the ratio between the long and short axes in a circle on the polymer which is deformed to an ellipse by the stretching process (92).

Tanizaki's model further assumes that at infinite stretching one specific axis in the molecular plane, the so-called orientation axis, is perfectly aligned with the stretching direction. The position of this axis is generally assumed to be in the molecular plane, although this has been questioned in one case (90). Tanizaki's orientation axis is not in general identical to the orientation axis ("long axis") defined by other authors (51).

An angle $\theta(\lambda)$ can now be defined:

$$2\cot^2\theta(\lambda) = R_{d\infty}(\lambda) \qquad (0^\circ \le \theta \le 90^\circ)$$

If the spectrum contains a purely polarized transition, i, at λ_i, the assumptions made imply that $\theta(\lambda_i)$ is the angle between the orientation axis and the transition moment for i.

For transitions of pure polarization in molecules of C_{2v} or D_{2h} symmetry, the minimum value of $\theta(\lambda)$, θ_{min}, is assumed to correspond to transitions along the molecular long axis (z), whereas the maximum value of $\theta(\lambda)$, θ_{max}, corresponds to in-plane short axis (y) polarization. In special cases θ_{max} is supposed to represent also out-of-plane (x) polarized transitions, which according to the authors cannot be separated from y-polarized (90). Since the orientation axis is assumed to be in the molecular plane, the following relation should hold:

$$\theta_z + \theta_y = 90^\circ$$

In a case with purely z- and y-polarized transitions this is identical to $\theta_{min} + \theta_{max} = 90^\circ$; at least when no x-polarized intensity is present, as discussed later. This relation is consistently used in the different applications of Tanizaki's model. It means that only one transition of pure polarization is needed for a determination of both θ_z and θ_y. The construction of reduced spectra $A_z(\lambda)$ and $A_y(\lambda)$ on the basis of Tanizaki's model has also been proposed (90):

$$A_z(\lambda)/A_y(\lambda) = (2-R_{d\infty}(\lambda)\cot^2\theta_z)/(R_{d\infty}(\lambda)-2\cot^2\theta_z) = r(\lambda)$$

$$A(\lambda) = A_z(\lambda) + A_y(\lambda) \tag{III,14}$$

where A is measured in an unstretched film. The equations, when correct, make a determination of $A_z(\lambda)$ and $A_y(\lambda)$ possible:

$$A_z(\lambda) = A(\lambda) \cdot (A_z(\lambda)/A_y(\lambda))/(1+A_z(\lambda)/A_y(\lambda)) = A(\lambda)\cdot r(\lambda)/(1+r(\lambda))$$

$$A_y(\lambda) = A(\lambda) - A_z(\lambda) \tag{III,15}$$

It is further claimed (90) that $A_x(\lambda)$ automatically is included in $A_y(\lambda)$.

In Tanizaki's work it is mentioned that for molecules of symmetry lower than C_{2v}, $\theta(\lambda)$ provides information about the position of the transition moment relative to the orientation axis. Similarly as discussed before, the relative sign of $\theta(\lambda)$ for a specific transition may be determined e.g. from studies of other molecules with similar chromophores, but different shapes.

In the analysis of the model it should be kept in mind that even a physically incorrect orientation model may provide correct results (correctly reduced spectra) at least for certain groups of molecules and specific experiments (e.g., although the common application of the Fraser-Beer model is based on a physically incorrect assumption about the orientation distribution, it gives correct results for e.g. one-photon processes in rod-shaped molecules). In the following,

Tanizaki's model, being considerably more complicated and less transparent than the Fraser-Beer model, will be analyzed (51) in terms of the quantities K_z, K_y and K_x corresponding to the observed spectra, and K_z^∞, K_y^∞ and K_x^∞ which would be obtained in the hypothetical situation of infinite stretching given by (III,12). The following relations hold for i = x, y, z:

$$R_{d\infty}(\lambda_i) = 2\cot^2\theta_i = 2K_i^\infty/(1-K_i^\infty) \qquad (III,16)$$

$$R_d(\lambda_i) = 2K_i/(1-K_i) \qquad (III,17)$$

$$K_i^\infty = (T-1+2K_i)/(3T-1) \qquad (III,18)$$

According to the model, $K_x^\infty = 0$ should hold for all samples. From (III, 18) it is seen that this will only be the case if

$$K_x = (1-T)/2$$

or equivalently since $K_x = 1-K_z-K_y$:

$$K_y = -K_z + (1+T)/2 \qquad (III,19)$$

In other words: the predicted in-plane transition moment directions will be correct if the observed real orientation distribution corresponds to a point $(K_z, -K_z + (1+T)/2)$ on the line $K_y = -K_z+(1+T)/2$ in the orientation triangle. This line is determined through T by the stretch ratio R_s, see (III,13).

When the condition (III,19) is fulfilled, the reduced spectra determined by (III,15) will be correct if $A_x(\lambda) = 0$, or if $\theta_z = 0$. If this is not the case, the expressions (III,14), (III,15), (III,4), and (III,16) give:

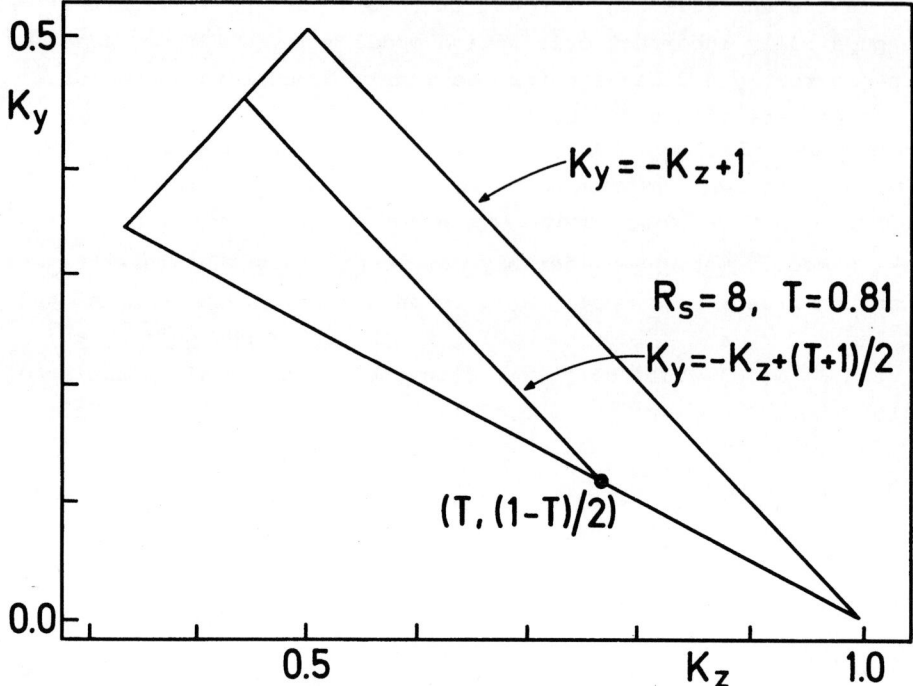

$$A(\lambda) \cdot r(\lambda)/[1+r(\lambda)] =$$

$$A_z(\lambda) - A_x(\lambda)/(\cot^2\theta_z - 1) \qquad\qquad (III,20)$$

which means that the procedure which should lead to $A_z(\lambda)$ instead will produce $A_z(\lambda)$ plus a (negative, since $\theta_z < 45^\circ$) contribution proportional to $A_x(\lambda)$. In a case where $A_z(\lambda) = 0$, $A_x(\lambda) > 0$, this would lead to negative results. Attempts to calculate $A_y(\lambda)$ from (III,15) will similarly lead to:

$$A(\lambda) \cdot (1-r(\lambda)/[1+r(\lambda)]) = A_y(\lambda) + A_x(\lambda) \cdot [1 + 1/(\cot^2\theta_z - 1)] \qquad (III,21)$$

which means that instead of $A_y(\lambda)$, or $A_y(\lambda) + A_x(\lambda)$, a weighted mixture of $A_y(\lambda)$ and $A_x(\lambda)$ is obtained, where the latter appears with the higher weight since $\cot^2\theta_z > 1$. The correct results for $A_z(\lambda)$ and $A_y(\lambda)$ are only found if $A_x(\lambda) = 0$ or if $\theta_z = 0$. The value $\theta_z = 0$ is obtained for a rod-like molecule with perfect extrapolation to infinite stretching. In all other cases the reduction method is a possible source of error, also when the model otherwise works, that is when $K_y = -K_z +$ (1+T)/2.

For molecules of C_{2v} or D_{2h} symmetry with regions of both pure z- and pure y-polarized intensity and no x-polarized intensity, the basic relation $\theta_{min} + \theta_{max} = 90°$ must hold according to Tanizaki's model. This is frequently not the case in the real spectra. The deviations are often significant, but are rarely seen as results of failures of the model. Instead they have been explained as due to (51):

1) Experimental errors, e.g. for acridine and phenazine (93) or 1,10-phenanthroline (94). In these works the reduced spectra shown are probably correct, but cannot possibly have been obtained directly by means of (III,14) and (III,15) as indicated in the text.

2) Deviations from expected symmetry, e.g. in the case of tropolone (95) where $\theta_{max} + \theta_{min} = 99°$ is taken as proof of inequivalency between the two oxygen atoms (deviation from C_{2v}-symmetry) in the negative ion:

Since the observation of $99°$ instead of the expected $90°$ is considered as a result of transition moment directions different from those possible in a molecule with C_{2v} symmetry, it is concluded that the molecule has symmetry lower than C_{2v} due to a difference between the two C-O bonds.

3) Mixed polarization of all y- or z-polarized bands ($\theta_z < \theta_{min}$, $\theta_y > \theta_{max}$, or both), as in the C_{2v} molecules 1,4-dihydroxynaphthalene (96), or 1,3-diazaazulene (90). In the latter the strong 40000 cm^{-1} band is predicted to have more than 30% contribution from short-axis polarized transitions. In doubly protonated 2,7-diaminofluorene (97), the sharp 33000 cm^{-1} peak is considered to be of strongly mixed polarization. Other examples of assumptions of mixed polarization are found for a number of molecules: phenanthrene (98), fluorenone (98),

fluorene (99), 1,4-diaminoanthraquinone (100), 1,4,5,8-tetraamino-anthraquinone (100) and 9-methylanthracene (101).

4) New, previously unknown transitions, of $\pi-\pi^*$ or $n-\pi^*$ type. Transitions of the former type are proposed in pyrene (102), where otherwise well understood spectral regions of dominating long axis polarization (Fig. 1) are clamined to contain short axis polarized transitions at wavelengths identical to those for the long axis polarized peaks. Transitions of $n-\pi^*$ type are proposed in 1,3-diazaazulene (90) and in 2-ethylthio-1,3-diazaazulene (103). In both these cases, out-of-plane polarized ($n-\pi^*$) absorption is claimed to be present with considerable intensity in the "$A_y(\lambda)$" curve, and the implicit assumption is made that x- and y-polarized intensity appear with equal weight. Also, the observation of intensity due to symmetry forbidden $n-\pi^*$ transitions is discussed (90, 104). Generally, the observed intensities seem very high compared with $n-\pi^*$ transition probabilities observed in other molecules.

The above mentioned cases are all characterized by the relation $\theta_{max} + \theta_{min} \neq 90^{\circ}$, which has been accounted for in different ways. In all cases the deviation might also be explained as failures in the description based on Tanizaki's model: The actual orientation distribution function does not correspond to the values $(K_z, K_y) = (K_z, -K_z +(1+T)/2)$. The conclusions mentioned in connection with 2) and 4) are hardly acceptable and those discussed under 1) and 3) are in most cases quite doubtful.

The suggestion (90, 103) that out-of-plane (x-) polarized intensity appears in the reduced spectra like the y-polarized intensity seems to be in disagreement with the basic assumption $\theta_z + \theta_y = 90^{\circ}$, except if the orientation distribution is supposed to be rod-like: $(K_z, K_y) = (K_z, (1-K_z)/2) = (T, (1-T)/2)$ according to (III,19). However, in a rod-like case (III,18) shows that $K_z^{\infty} = 1$, $K_y^{\infty} = K_x^{\infty} = 0$. These values correspond to $\theta_z = 0^{\circ}$, $\theta_y = 90^{\circ}$, which is in disagreement with the values obtained in refs. (90, 103). Thus, it is clear that possible out-of-plane polarized intensity will appear in the reduced spectra also according to (III,20), and not as a simple contribution to the y-polarized curve.

It should be emphasized that Tanizaki's model has worked in many other cases as a correct, but probably not very efficient, method for the treatment of the excellent experimental data obtained by Tanizaki and coworkers using stretched PVA sheets. Such examples are found in Refs. (105, 106). However, even in these cases there seems to

be no reason to prefer the Tanizaki model over the more general description given above (6, 19, 24, 51, 54, 55).

An interesting example of a combination of Tanizaki's model with
the Fraser-Beer model is found in an investigation of nucleic acid
bases oriented in stretched PVA (107). In this treatment the molecules
are supposed to be aligned so that $K_y = -K_z + (1+T)/2$ as required by
Tanizaki's model, and at the same time to fulfill the relation $K_y =$
$(1-K_z)/2$ for a rod-like distribution. Thus again $(K_z, K_y) = (T, 1-T/2)$.
The authors show that the value of the Fraser-Beer parameter f is
(Eq. (III,10)):

$$f = \frac{(R_d-1)(R_{d\infty}+2)}{(R_{d\infty}-1)(R_d+2)} ,$$

where both R_d and $R_{d\infty}$ are functions of the wavelength. This expression
is calculated for each molecule and is found to be constant to 2%
over the whole spectral region. This is taken as proof of the uniqueness
of the orientation axis, which in Tanizaki's model is an obvious problem
for low symmetry molecules, and as support for the models used. However,
if the relation (III,12) is inserted in the above expression for f, the
result is: $f = (3T-1)/2$, which necessarily must be constant for any
given stretch ratio and provides little support for either of the
orientation models used. A very recent discussion of Tanizaki's model
applied to rod-shaped molecules has been given by Nordén (108), who
proposes some modifications and mathematical simplifications.

Tanizaki's method has earlier been criticized by Popov (109),
who suggests that for small stretch ratios the model implies that all
angles θ will necessarily be close to 45°. It is not clear whether
this critique takes into account the extrapolation to infinite stretching. A more obvious basis for critique of the model is the application
to molecules with two or three identical orientation constants (rod-
shaped, spherical or disc-shaped molecules). In the first case Tanizaki's
model requires $K_z = T = (1-K_y)/2$, in the last $K_z = (T+1)/4 = K_y$, where
T and the K's only depend on the stretch ratio R_s, according to (III,13).
The same relations must hold for all molecules of a given type, even
when their size is quite different (e.g. benzene and coronene). Moreover,
two almost spherical molecules, one slightly rod-shaped and one slightly disc-shaped would for normal T-values (0.5-0.9) be aligned completely differently. This consequence of Tanizaki's model is obviously
physically unacceptable.

6c. Other assumptions of orientation distributions.

For many years, two of the pioneers in the field of LD spectroscopy, Popov and Smirnov, preferred to consider the information obtained
from stretched sheet spectra as qualitative information (110-13). In
1970 Popov proposed an orientation model (114-16) that made a quantitative determination possible of the angle ϕ_i between the orientation
axis and the transition moment for the i'th transition:

$$\cos^2\phi(\lambda_i) = [(3A-1) + g(\lambda_i)(3-A)]/(3A-1)(3-g(\lambda_i))$$

where $g(\lambda_i)$ is the ratio $[E_Z(\lambda_i)-E_Y(\lambda_i)]/[E_Z(\lambda_i)+E_Y(\lambda_i)]$ for the transition at λ_i and A is an orientation constant for a given molecule and
degree of stretching. However, for a pure long axis polarized transi-
tion: $\cos^2\phi(\lambda_z) = 1$, $g(\lambda_z) = (3K_z-1)/(K_z+1)$, and for a pure y-polarized
transition: $\cos^2\phi(\lambda_y) = 0$, $g(\lambda_y) = (3K_y-1)/(K_y+1)$. Thus, $K_z = A$ and
$K_y = (1-A)/2 = (1-K_z)/2$, which shows that the model essentially
corresponds to the assumption of a rod-like orientation distribution:
$K_y = (1-K_z)/2$.

A few years ago Popov stated that one-parameter models "are
inadequate for the description of the orientation distribution of
planar molecules having a shape with little elongation" (117). This is
identical to the suggestions made by Eggers and Thulstrup (19) much
earlier and is contained in the general description given above. Popov
proposed a new model, which essentially is an elaboration of the model
by Eggers, Michl and Thulstrup (6, 19, 24, 51, 54, 55), but the
applicability seems to be limited to planar molecules. In the new model
three parameters are used. One parameter, A, defined as the average of
the cosine square of the angle θ between the stretching direction and
the chains in the polymer after stretching, represents an aspect not
included in the general description given earlier. A is used as an
empirical parameter in contrast to the quantity T in Tanizaki's model,
which in some respects serves a similar purpose.

The remaining two parameters, B and C, are connected with the
solute molecule orientation and serve a purpose equivalent to that of
K_z and K_y. In their definition it is first assumed that the solute
orientation distribution has cylindrical symmetry around each polymer
chain. It must be added that this assumption seems hard to justify
unless the polymer chains are widely separated. The second parameter B
is defined as the average cosine square of the angle α between the
orientation axis and the projection of the polymer chain direction
into the molecular plane. The third parameter C is the average cosine
square of the angle δ between the polymer chain direction and its

projection in the molecular plane. It is further assumed that the distributions of α and δ are independent. The relations between (A,B,C) and (K_z,K_y) are:

$$B = (2K_z+A-1)/2(K_z+K_y+A-1)$$

$$C = 2(K_z+K_y+A-1)/(3A-1) \qquad\qquad (III,22)$$

$$2K_z = (3A-1)\cdot B\cdot C + 1-A$$

$$2K_y = (3A-1)(1-B)C + 1-A \qquad\qquad (III,23)$$

In Popov's model the determination of B and C can be made for a symmetrical molecule with purely y- and z-polarized transitions, whereas A can be determined from studies of the pure polymer. When all three parameters are known, the angle $\phi(\lambda_i)$ between the transition moment for the i'th in-plane transition at λ_i and the orientation axis can be determined from the expression (117):

$$g_i(\lambda_i) = (E_Z(\lambda_i)-E_Y(\lambda_i))/(E_Z(\lambda_i)+E_Y(\lambda_i)) =$$

$$[(3A-1)\{(3C(1-B)-1) + 3C(2B-1)\cos^2\phi(\lambda_i)\}]/[(3-A) + C(3A-1)(1-B) +$$

$$C(3A-1)(2B-1)\cos^2\phi(\lambda_i)] \qquad\qquad (III,24)$$

If Popov's model is analyzed in terms of the orientation constants (K_z, K_y) it is clear from $(III,23)$ that a number of points inside the orientation triangle can no longer be reached unless the polymer chains are perfectly aligned $(A=1)$. That is, the limit $K_z+K_y \leq 1$ has been replaced by $K_z + K_y \leq (1+A)/2$. Similarly as for (K_z,K_y) in the orientation triangle a figure illustrating the possible values of (B,C) may be drawn (51). The result is a distorted triangle, since from $(III,22)$: $\frac{1}{2} \leq B \leq 2-1/C$, $C \leq 1$. Each side in this distorted triangle corresponds to a side in the smaller allowed orientation triangle for (K_z, K_y).

The main potential advantage of Popov's model compared with other models is the possibility that the constants (B,C) may be independent of the type of polymer and degree of stretching used, whereas e.g. (K_z,K_y) for a given molecule depend on the choice of polymer as well as on the stretch ratio. However, so far it has not been shown that this aspect of the model is useful in practice. On the contrary, one of the basic assumptions of the model seems to give reason for

concern. This is the assumption of rotational symmetry of the distribution of the solute molecules around each polymer chain, which implies
that the highest degree of alignment of the solute molecules is that
of the polymer chains. This limitation is in disagreement with some
experiments (69), in which a better alignment of solute molecules than
of the polymer chains has been observed. Finally, the Popov model seems
to lack the simplicity and applicability of the earlier more general
description by Eggers, Michl and Thulstrup.

Fig. 2 clearly shows that the solute alignment is related to
other molecular properties. Suggestions by Popov for a relation between
molecular shape and orientation (118, 119) which is essential for his
model, seem to be much too simple. For example it is predicted that all
molecules with the same length and width, such as pyrene and perylene,
will have the same orientation factors corresponding to (K_y, K_z). This
is clearly not the case (Fig. 2). On the other hand, shape does seem
to be a determining factor for the orientation, although other
relations have been proposed. For example did a theoretical investigation of the dispersion interaction for a solute in a stretched sheet
lead to a relation between K_z and the molecular anisotropy (120).
However, a more recent study of orientation mechanisms (121) finds
this relation unrealistic.

Other attempts to define orientation models have been made.
One interesting model is that by Gangakhedkar et al. (122). The authors
assume (implicitly) a rod-like orientation distribution function, but
no assumption is made about the position of the orientation axis. This
position is found from observations of the IR dichroic ratios for three
normal modes in the solute. Since such information usually is not
available, the model seems to have very limited applicability.

6d. Conclusions based on incomplete information.

In the LD literature there are numerous examples of incorrect
conclusions about transition moment directions in planar molecules
caused by the investigator's inability to extract the full information
from the observed spectra. Even recently (123), investigations have
been reported in which only the experimental curves are studied for
molecules of C_{2v} and D_{2h} symmetry, in spite of the fact that the reduced
curves reveal much more information, as pointed out many years ago
(6,19). In other cases, dichroic ratios like (E_z/E_y) or $(E_z-E_y)/(E_z+E_y)$
have been determined and the shape of a single curve of this type
studied. This clearly decreases the possibilities for a complete
understanding of the spectrum, compared with the study of a set of

(correctly) reduced spectra. An example of an incorrect prediction of
the position of a transition due to incomplete extraction of information
is found for anthracene (21).

The use of a simple dichroic ratio may lead to several mis-
conceptions, e.g. that $(E_Z/E_Y) < 1$, or equivalently $E_Z > E_Y$, for a
short axis (y) in-plane purely polarized transition. In the case of a
y-polarized spectral region $(E_Z/E_Y) < 1$ is identical to the statement
$K_y < 1/3$, which according to Fig. 2 is not fulfilled for all molecules.
However, $(E_Z/E_Y) > 1$ is always true for a long axis (z) purely polarized
transition and $(E_Z/E_Y) < 1$ always fulfilled for a short axis (x) out-of-
plane polarized spectral region.

In spite of many examples of unfortunate attempts to describe
the orientation distribution functions and to evaluate the spectra, a
large amount of high quality work is carried out today in the field of
stretched sheet visible - UV LD spectroscopy. Interesting research
based on modern evaluation methods for the spectra obtained in stretched polymers is carried out in many countries, such as Germany (124),
Sweden (15), Australia (121) and USA (31), and excellent experimental
work is performed in Japan (125), Israel (82), the Netherlands (48),
and the Sovjet Union (118, 119). In addition important results are
obtained with other solvents, such as liquid crystals (41), and many
new research groups, especially among biologists, are taking up
research in LD spectroscopy. For these, the existing literature in
the field, particularly on the evaluation of experimental data, may
easily seem confusing, in some cases even discouraging. In view of
the increasing interest in new applications of LD spectroscopy, this
is very unfortunate and the need for standard textbooks on the subject
seems obvious. It may be added that reviewers of papers on LD spectroscopy, particularly for some of the leading chemistry journals, in some
cases seem to have overlooked serious mistakes and that several journals
must take part of the responsibility for the present problems in the
literature on LD spectroscopy.

7. Interpretation of LD data for low symmetry molecules.

The special problems connected with the evaluation of spectra
of molecules, for which any transition moment direction is theoretically
possible, has been mentioned in connection with the discussion in the
preceding sections. It was mentioned that while less accurate LD
data for purely polarized transitons in molecules of C_{2v} or D_{2h} symmetry
often allows a distinction between e.g. long-axis and short-axis
polarized spectral features, the determination of transition moment

directions in unsymmetrical molecules usually requires much more precise experimental information.

The absorption spectrum of a low symmetry molecule can be characterized by a set of absorption curves $\{A_i(\lambda)\}$ each corresponding to a specific transition i. The transition moment of i forms an angle (iz) with the orientation axis in the molecule, and angles (iy), (ix) with the two remaining axes. The special choice of axes z, y, and x, which ensures a diagonal K-matrix (e.g. $<\cos\beta\cos\beta'> = 0$), is no longer determined by symmetry alone, and an estimate of the position of the axes must be made (51). In this system of axes (x,y,z) the following relations hold (51, 55):

$$E_z(\lambda) = \sum_i [K_z \cos^2(iz) + K_y \cos^2(iy) + K_x \cos^2(ix)] A_i(\lambda)$$

$$E_y(\lambda) = \tfrac{1}{2} \sum_i [(1-K_z) \cos^2(iz) + (1-K_y) \cos^2(iy) + (1-K_x) \cos^2(ix)] A_i(\lambda) \quad (III,25)$$

Again, d_i is defined so that a spectral feature due to $A_i(\lambda)$ disappears from the linear combination:

$$E_z(\lambda) - d_i E_y(\lambda) \qquad\qquad (III,26)$$

A parameter K_i is defined by:

$$K_i = d_i/(d_i+2) = K_z \cos^2(iz) + K_y \cos^2(iy) + K_x \cos^2(ix) \qquad (III,27)$$

In order to determine the angles (iz), (iy) and (ix) the orientation constants K_z, K_y and K_x must be known. They may be obtained from assumptions about the moment directions of specific transitions in the molecule, or from comparison with (interpolation between) orientation constants for symmetrical molecules (Fig. 2) which are expected to align similarly. Assumptions of specific properties of the orientation distribution function (rod-like, disc-like) are also in some cases acceptable. Even when the K's are known, (III,25-27) do in general not provide sufficient information for a determination of the size of (iz), (iy) or (ix). Such a determination is only possible in some special cases (51):

1) The transition moment directions are restricted to a plane ($\pi-\pi^*$ transitions in planar molecules), e.g. (ix) = 90° for all transitions i. Then from (III,25) and (III,27):

(ix) $= 90^{\circ}$

$$\tan^2(iz) = \cot^2(iy) = (K_z-K_i)/(K_i-K_y) \qquad\qquad (III,28)$$

An equivalent expression was given in (55), and later applications are numerous, e.g.ref.(8, 31, 43, 126). Examples are shown in Fig. 4 and Table 2, which illustrate the procedure and the results that can be obtained. In the case of low symmetry molecules, absolute angles (signs of (iz) and (iy)) may be determined from comparison with fluorescence polarization data (43) or from LD (stretched sheet) studies of molecules with identical chromophores, but with different orientation distribution functions.

2) The orientation distribution function is rod-like: $K_y = K_x = (1-K_z)/2$. Then from (III,25), (III,27) and $\cos^2(iz) + \cos^2(iy) + \cos^2(ix) = 1$:

$\underline{K_y = K_x}$

$$\tan^2(iz) = (K_z-K_i)/(K_i-K_y) \qquad\qquad (III,29)$$

Clearly, (iy) and (ix) cannot be determined separately for a rod-shaped molecule.

3) The orientation distribution function is disc-like: $K_y = K_z = (1-K_x)/2$. Then:

$\underline{K_y = K_z}$

$$\tan^2(ix) = (K_i-K_x)/(K_z-K_i) \qquad\qquad (III,30)$$

In this case (iz) and (iy) are undetermined.

A determination in the general case of the numerical values of all three angles might be possible from at least three linear independent measurements, as discussed in connection with the description of symmetrical molecules. Such a determination would in general require very accurate experimental data.

8. Identical chromophores in different molecules.

In some molecules all visible and near UV absorptions can be seen as a result of changes in the electronic wavefunctions which involve only part of the molecule; usually this part is the π-electron

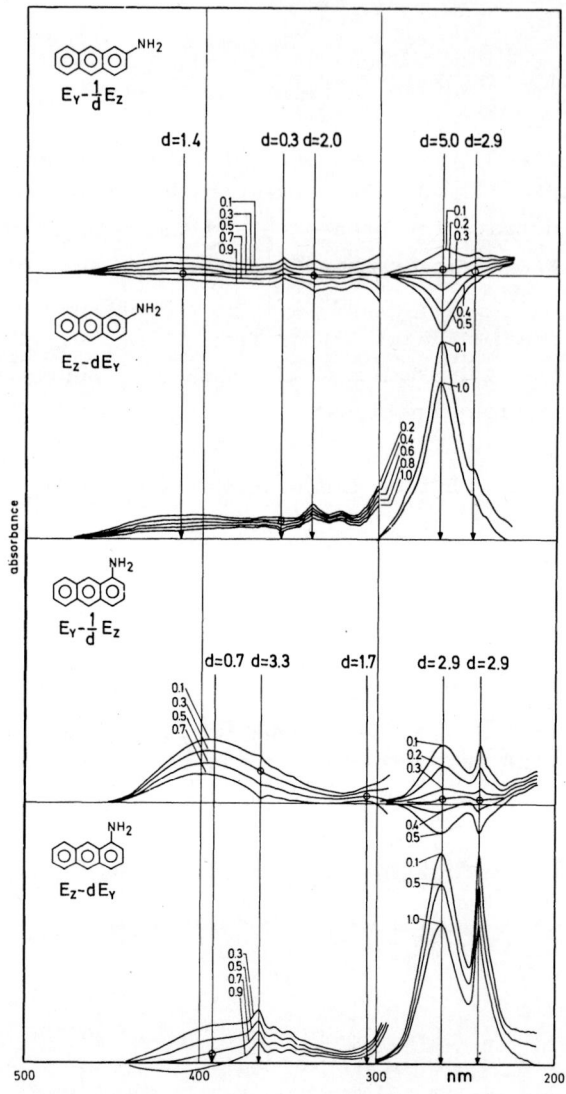

Fig. 4. 1-Aminoanthracene (bottom) and 2-aminoanthracene (top) (8).
Linear combinations of E_Z and E_Y and values for d_i corresponding to
important spectral features are shown. These are used for a determination of transition moment directions (Table 2).

Table 2

Transition Moment Directions in Aminoanthracenes (8)

	E(ev)		ϕ	
	Exp.	Calc.	Exp.	Calc.
9-NH$_2$	3.0	3.1	90	90
	3.3	3.5	0	0
	-	3.9	-	0
	4.5	4.8	90	90
	4.6	4.8	0	0
	4.9	5.0	0	0
1-NH$_2$	3.2	3.0	+88 ± 10	+80
	3.4	3.6	0 ± 10 or +16 ± 10	+20
	-	4.1	-	-56
	4.0	4.5	+42 ± 10 or -26 ± 10	+31
	4.7	4.5	0 ± 10	-10
	5.1	5.1	+16 ± 10	0
	5.6	5.3	+72 ± 10 or -56 ± 10	-47
2-NH$_2$	3.0	3.1	-57 ± 5	-58
	3.5	3.6	+82 ± 8	+86
	4.4	4.3	+34 ± 10 or -24 ± 10	+17
	-	4.5	-	+35
	4.7	4.8	- 3 ± 10	+14
	5.0	4.8	-21 ± 5	+ 1

A comparison between experimental and calculated transition moment
directions in 3 aminoanthracenes (8). The experimental values were
evaluated by means of (III,28). In the table, E is the transition energy
and ϕ is the angle of the transition moment with the long axis of the
anthracene chromophore; the angle between this axis and the orientation
axis was assumed to be 8° corresponding to the direction perpendicular
to the smallest molecular cross-section. Some signs (measured counter-
clockwise)* for ϕ could be obtained from comparison with fluorescence
polarization data (K. Rotkiewicz and Z.R. Grabowski, Trans. Faraday
Soc. 65, 3263 (1969)). The calculated results were obtained from PPP-
calculations (8).
*) See Fig. 4.

chromophore. Such molecules may in several cases be treated in a special way in connection with the interpretation of their spectra.

If two or more different molecules have identical chromophores and almost identical absorption spectra, but different geometries, additional information on either transition moment directions or on orientational properties (the K's) can be obtained from a comparison between the LD spectra of the compounds. In particular, it may be possible by inspection to determine the otherwise unknown signs of transition moment angles in low symmetry molecules. A general prescription for such applications is not possible since intensities and vibronic structure in the absorption spectra usually differ considerably between the molecules in spite of their identical chromophores. Therefore, a reduction procedure involving spectra of different molecules will hardly be possible except in very special cases.

A low symmetry molecule with a C_{2v} and D_{2h} chromophore may frequently be treated as a symmetrical molecule; as examples can be mentioned methyl and fluoro derivatives of anthracene (8). In this connection, it should be noted that changes in transition moment directions in symmetrical chromophores in low symmetry molecules are likely to be much larger for weak than for intense electronic transitions.

In some cases the "long" axis (z) in one molecule may correspond to e.g. the second axis (y) in another molecule containing the same chromophore and having almost the same absorption spectrum. This has been observed in the spectra of acenaphthylene and 1,2-dibromoacenaphthylene (64):

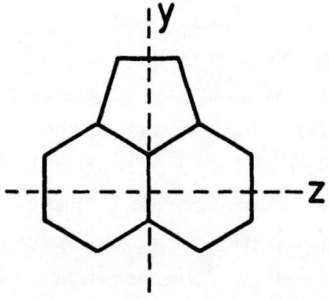

$(K_z, K_y) = (0.43, 0.38)$ $(K_z, K_y) = (0.54, 0.33)$

9. Non-uniaxial samples.

In some of the work (6, 7, 19, 61) prior to the earlier mentioned x-ray diffraction investigations (55) of stretched polyethylene, no assumption of uniaxial symmetry was made for the samples. This did not impose serious restrictions on the extraction of detailed information from the observed spectra. For instance, it was shown that reduced spectra for symmetrical molecules with $A_x(\lambda) = 0$ could be obtained. For any sample of molecules with $A_x(\lambda) = 0$ one has:

$$E_z(\lambda) = K_z A_z(\lambda) + K_y A_y(\lambda)$$

$$E_y(\lambda) = K_z' A_z(\lambda) + K_y' A_y(\lambda) \qquad (III,31)$$

Obviously, the quantities K_z/K_z' and K_y/K_y' may be determined from (III,31) when z- and y-polarized spectral features can be identified in the spectrum, since:

$$K_i/K_i' = d_i, \quad i = y, z$$

From (III,31) it is clear that the reduced spectra $A_z(\lambda)$ and $A_y(\lambda)$ can be written (6, 19):

$$A_z(\lambda) = C_z[E_z(\lambda) - (K_y/K_y')E_y(\lambda)]$$

$$A_y(\lambda) = C_y[E_y(\lambda) - (K_z'/K_z)E_z(\lambda)] \qquad (III,32)$$

where C_z and C_y are constants, which depend on K_z, K_z', K_y, and K_y'. Their size cannot be determined from the observed spectra $E_z(\lambda)$ and $E_y(\lambda)$. However, their relative size may be determined by comparison with a spectrum of a sample with random orientation $A(\lambda)$, since

$$A(\lambda) = C[A_z(\lambda) + A_y(\lambda)].$$

where C is an unknown constant.

It is a condition that the spectrum $A(\lambda)$ is similar to $E_z(\lambda)$ and $E_y(\lambda)$ with respect to spectral resolution, solvent effects, etc. in order for this method for the determination of C_z/C_y to be correct. Fortunately, the determination of C_z/C_y is generally not of great importance; usually the two curves proportional to $A_z(\lambda)$ and $A_y(\lambda)$ which can be obtained according to (III,32) are sufficient for the interpretation of the spectrum (6, 7, 19, 61).

10. Interpretation of MCD-spectra.

In the introduction it was mentioned that all molecules show an MCD effect and that this effect for molecules without degenerate electronic states is described by the non-vanishing MCD B-terms, one for each electronic transition. The MCD of a given transition may be positive or negative, and contributions of different signs may cancel in regions of overlapping transitions.

In the following it is assumed that the observed MCD spectra can be accounted for by the B-terms only. Thus A- and C-terms as well as the natural CD effect are supposed to be zero or to have been removed from the spectra for the molecules investigated. Under these circumstances, the interpretation of MCD spectra of isotropic samples is usually reasonably straightforward at least when the absorption bands do not overlap badly.

A very exciting possibility is the measurement of MCD of oriented samples. The theory for the orientational dependence of MCD is quite complicated, but the result expressed in K_x, K_y and K_z is not more complicated than that for ordinary absorption (11b). However, since no experimental results are yet available, these aspects will not be discussed in more detail here.

The observed MCD of isotropic samples may be expressed in units of molar ellipticity per unit magnetic field ($[\theta]_M$ in deg. $m^{-1} \cdot \ell \cdot mol^{-1} \cdot gauss^{-1}$), and is related to the calculated B-terms (in units of $10^{-3} \beta_e D^2/cm^{-1}$) by the relation (9):

$$B(i \rightarrow j) = -33.53^{-1} \cdot \int \theta_M(i \rightarrow j) \, d\nu/\nu$$

where ν is the wavenumber, and the integration is carried out over the spectral region corresponding to the transition $\psi_i \rightarrow \psi_j$. All other contributions are supposed to have been removed. In MCD spectra neighboring transitions of different polarization may mix strongly according to the expression for $B(0 \rightarrow F)$ given in the next chapter. This means that the B-terms for neighboring transitions often have different signs, and cancellation effects may thus frequently occur in connection with closelying transitions. If such overlap is not recognized, it will often lead to too low numerical experimental values for the MCD compared with the theoretical values.

IV. THEORETICAL MODELS

Already in the title of this paper it has been assumed that the
molecules which will be treated have at least a plane as symmetry
element; this means that the molecular orbitals and electronic states
can be classified according to the irreducible representations of the
C_s point group. It should be noted, however, that also molecules with
only approximate C_s symmetry may be successfully described by the
models discussed below, since the part of the nuclear potential that
is not invariant under the operations of the group C_s often represents
only a minor perturbation.

1. The π-electron approximation.

A significant simplification of the computational work involved
in studies of the electronic structure of planar molecules can be
obtained by assuming the π-electron approximation. This approximation
has been extensively discussed by Lykos and Parr (127) and more recently
by Linderberg and Öhrn (128). It is assumed that the electronic
Hamiltonian commutes with the number operator for electrons in the σ-
and π-field, respectively, where the electron field operator $\psi(x)$ has
been expressed as a sum of components $\psi_\sigma(x)$ and $\psi_\pi(x)$ corresponding to
the totally and non-totally symmetric irreducible representations of
C_s. Furthermore, the operator nature of $\psi_\sigma(x)$ is neglected and average
values are used for the potential of the σ-field. This corresponds to
writing the electronic wavefunctions as antisymmetrized products of the
antisymmetrized functions Σ and Π, where e.g. Π is a function of the
coordinates of a fixed number, n_π, of so-called π electrons and Σ
is a function (the same for all electronic states) of the coordinates
of n_σ σ-electrons.

The π-electron approximation is usually quite satisfactory for
a description of the visible and long wavelength UV spectra of conjugated molecules. In general, transitions in these molecules which
correspond to large changes of Σ have high energy or low intensity. One
of the most important groups of such transitions in the low energy
region is the so-called n-π* transitions, which correspond to excitation
of an electron from a non-bonding (σ-type) spin orbital, which in some
cases (e.g. aza nitrogen) have quite high energy, to a vacant π spin-
orbital (π*). Allowed n-π* transitions are polarized perpendicular to
the molecular plane; they have in general low intensity and are often
difficult to observe in regions with symmetry-allowed π-π* absorption.

A more difficult problem in the π-electron approximation than
the neglect of excitations involving σ-orbitals is the assumption of

an unchanged Σ function for all states. This assumption is reasonable
if the polarizability of the σ-skeleton is small (128) but experimental
investigations on graphite layers by Taft and Phillip (129) have
shown that this is not always the case. The effect of the σ-polariza-
bility corresponds to a reduction of the electronic interaction within
the π-system, and the empirical values (130) - determined from observed
transition energies - for the one-center and nearest neighbor electron
repulsion integrals are considerably lower than the values calculated
from e.g. Slater orbitals. Unfortunately, this problem cannot be solved
simply by introducing a dielectric factor in the electron repulsion
integrals (128).

2. Semiempirical calculations.

The use of empirical values for atomic integrals together with
the assumption of zero differential overlap, ZDO (131), has drastically
reduced the computational work involved in the calculation of spectra
of π-electron systems. The model most commonly used is that of Pariser
and Parr, and Pople (132). The Hamiltonian in this so-called PPP model
can be written (128):

$$H = H_0 + \sum_{r,\nu} \alpha_r n_{r\nu} + \sum_{rs,\nu}' \beta_{rs} a_{r\nu}^+ a_{s\nu} + \frac{1}{2} \sum_{rs,\nu\nu'}' \gamma_{rs} n_{r\nu} n_{s\nu'} \qquad (IV,1)$$

where H_0 represents the nuclear repulsion, $a_{r\nu}$ is the annihilation
operator for electrons in π orbital r with spin ν, $n_{r\nu} = a_{r\nu}^+ a_{r\nu}$ is the
number operator for spin orbital $(r\nu)$, α_r and β_{rs} are diagonal and out-
of-diagonal matrix elements of the core Hamiltonian between spin
orbitals $u_{r\nu}$ and $u_{s\nu}$, and γ_{rs} is the electron repulsion integral
between the densities $|u_{r\nu}|^2$ and $|u_{s\nu}|^2$. The primes on the summation
signs indicate omission of terms where the operators would correspond
to the same spin orbital. α_r, β_{rs} and γ_{rs} are (empirical) parameters
of the model; the value of α_r does not influence the results for
transitions in hydrocarbons and $\beta_{rs} = 0$ whenever r and s do not
correspond to orbitals on neighboring atoms. The exact solution of the
model corresponds to the full CI [CCI(133)] but usually only an
approximate solution is found, such as singly excited CI [SCI(133)] or
Time Dependent Hartree-Fock [TDHF (134)]. The values of the parameters
depend on the approximation used. A suitable set of parameters for SCI
is (135):

$$\beta_{\mu\nu} = -2.32 \text{ eV}$$

$$\gamma_{\mu\nu} = 14.4 \text{ eV}/[1.33 + R_{\mu\nu}(\text{Å})]$$

When more configurations are included the use of more negative values of $\beta_{\mu\nu}$ and a less steep distance dependence in $\gamma_{\mu\nu}$ gives the best agreement with experiment.

The theoretical determination of electronic transitions has been done both directly, by means of the Time Dependent Hartree-Fock method (TDHF) (136, 134), for a review see (137), or from wavefunctions and energies for the ground and excited states, calculated in a CI scheme. The latter method, which has been widely used, has one drawback: The results for electronic transition moments $\vec{M}$ depend on which of the expressions (dipole lengths and dipole velocity) for $\vec{M}$ (138) that is being used. Linderberg (139) has shown that in ZDO-models the commutation relation between $\vec{r}$ and H: $i\frac{d\vec{r}}{dt} = [\vec{r}, H]$, can be secured by using the following expression for the matrix elements between AO's of the momentum operator:

$$<\mu|\nabla|\nu> = (m/\hbar^2) \, \beta_{\mu\nu}(\vec{R}_{\mu} - \vec{R}_{\nu}) \,.$$

When such values of $<\mu|\nabla|\nu>$ are used, the two formulas for the transition moment will lead to the same result for an exact solution of the model (CCI). Also TDHF will yield identical results from the two expressions, while a CI scheme short of CCI in general will lead to different results. This is not very satisfactory, since both expressions seem theoretically justified. However, in most cases the two calculated results are close compared with the difference between calculated and experimental results, and in practice the deviations may usually be considered acceptable. A study (133) has shown that a careful choice of a limited number of configurations is able to give results that are in good agreement with respect to the two expressions for the transition moment.

3. Calculation of MCD B-terms. Alternant hydrocarbons.

For the calculation of MCD B-terms, which for molecules without degenerate electronic states are the only important MCD-terms, the following perturbation expression for B(0→F), corresponding to the transition from the (ground) state |0> to an excited electronic state |F>, is usually used (140):

$$B(0 \rightarrow F) = \sum_{I \neq 0} \langle I|-i\vec{M}|0\rangle \cdot \langle 0|\vec{M}|F\rangle \times \langle F|\vec{M}|I\rangle \cdot (E_I - E_0)^{-1}$$

$$+ \sum_{I \neq F} \langle F|-i\vec{M}|I\rangle \cdot \langle 0|\vec{M}|F\rangle \times \langle I|\vec{M}|0\rangle \cdot (E_I - E_F)^{-1} \qquad (IV,2)$$

where the sums are over electronic states $|F\rangle$ with energies E_F, and $\vec{M}$ is the magnetic and $\vec{M}$ the electric dipole moment operator. One problem is that the results in general will be origin dependent for molecules with less than D_{2h} symmetry. In the case of D_{2h}, there is no origin dependence since in general changes of origin perpendicular to symmetry axes do not affect the result. When the expression by Linderberg is used for the calculation of $\vec{M}$ and $\vec{M}$ (see later) the origin dependence disappears for the exact solution (CCI). This solution is often not possible for larger molecules and a choice of origin in the molecular plane within the framework of the molecule usually provides reasonable results for MCD. It is possible to avoid the origin dependence problems by using gauge-invariant (London) orbitals (141); unfortunately this increases the computational work considerably. The importance of the use of gauge-invariant orbitals has been discussed by several authors (142, 143).

Special problems are encountered in connection with the calculation of transition moments and MCD B-terms for the important group of alternant hydrocarbon molecules. In ZDO-models, the description of the electronic states of these molecules contains a special pairing symmetry, and the states can be classified as plus (+) or minus (-) states (144, 145). Matrix elements of both the magnetic and electric dipole moment operator between states of the same alternant pairing symmetry are predicted to be zero:

$$\langle -|\vec{M}|-\rangle = \langle +|\vec{M}|+\rangle = 0; \quad \langle -|\vec{M}|-\rangle = \langle +|\vec{M}|+\rangle = 0$$

which means that not only is the calculated oscillator strength zero, but the direction is undetermined. Even worse, when the above relations are inserted in the expression (IV,2) all B-terms vanish. This does not agree with experimental observations: transitions from a (-) ground state to excited (-) states are usually fairly weak, but often easily observable, and the MCD spectra of alternant hydrocarbons are often weak, at least in the region of the first few transitions, but they do contain important information, for which a theoretical description clearly would be valuable.

The problems connected with the description of alternant hydrocarbons have been attacked both by means of a more explicit

inclusion of the interaction between σ- and π-electrons (modified CNDO/S
(146)) and in a π-electron model by giving up the usual ZDO-approximation
and including overlap explicitly (147). The latter method was not
accepted without considerable hesitation, partly because of the generally
impressive success of ZDO models like PPP, and of the results of Gladney
(148), who showed that inclusion of overlap did not result in general
improvements in calculated energies for alternant hydrocarbons.

The explicit inclusion of overlap increases the computational
work considerably, and requires (for hydrocarbons) three additional
energy parameters, so-called penetration integrals, which describe
the penetration of π-electrons into the σ-cores. Consistency requires
inclusion of two-center penetration integrals between adjacent atoms
(including hydrogen), when overlap is considered explicitly. This makes
it necessary to consider penetration integrals of the types: $(\mu:\nu\nu)$,
$(H:\nu\nu)$ and $(\mu:\mu\nu)$, where the double index indicates a π charge (overlap)
density and the single index indicates a carbon (or hydrogen) atom core.
The values of these parameters were first taken from theoretical
considerations (148).

The results of the calculations (147) confirm the conclusion
of Gladney (148) that no obvious improvement of transition energies is
obtained by inclusion of overlap. However, the exact alternant pairing
symmetry is no longer present. This means that the former pairing
symmetry forbidden transitions become allowed, and the calculated MCD
B-terms do no longer vanish. In many compounds, however, the alternant
pairing symmetry, which according to experimental results is approximately fulfilled for most benzenoid hydrocarbons, is completely removed,
and matrix elements such as $<-|\vec{M}|->$ are strongly overestimated. This
leads to incorrect results for transition moments and especially for
MCD B-terms. The latter has to do with the fact that the too strong
breakdown of the alternant pairing symmetry causes contributions to
MCD from mixing of approximate (+) and (-) states to be overestimated.
Experiments indicate that this mixing often is of minor importance -
in cases like anthracene (149), where the two first transitions to
excited singlet states of approximate (-) and (+) symmetry are close
in energy, the relation $B(0\rightarrow(-))\sim-B(0\rightarrow(+))$ is far from fulfilled,
as it should be if the mixing between the two close-lysing states were
dominant.

The results also show that the strong deviations from perfect
pairing are accompanied by too high values of the π-electron charge
on carbon atoms with bonds to hydrogen atoms.

A different effect of the inclusion of overlap is the appearance

of non-zero matrix elements of the magnetic dipole moment operator
between non-neighbor atomic orbitals. These terms are zero in ZDO models
like PPP, since $\vec{M} = i\beta_e(\vec{R} \times \vec{V})$ where β_e is the Bohr magneton, and
according to Linderberg (139):

$$\langle\mu|\vec{V}|\nu\rangle \propto \beta_{\mu\nu}(\vec{R}_\mu - \vec{R}_\nu)$$

where $\beta_{\mu\nu} = 0$ when μ, ν are non-neighbors. In the calculations including overlap, the Linderberg relation is also used, but $\beta_{\mu\nu} = H_{\mu\nu}^{core}$
does no longer vanish between non-neighbor orbitals. This makes a
mixing of states with the same alternant pairing symmetry possible, also
when this symmetry is exact for the wavefunctions.

It has been demonstrated (147) that it is possible to adjust
the values of the penetration integrals so that a more even π-charge
distribution is obtained, by decreasing the value of the hydrogen
penetration integral (H:$\nu\nu$) and increasing the value of the carbon
penetration integral (μ:$\nu\nu$). At the same time approximate alternant
pairing symmetry is reestablished for most benzenoid hydrocarbons. In
such cases the neighbor interaction loses importance, and the terms
that correspond to non-neighbor matrix elements of H^{core} (or of $\vec{M}$)
become dominant in the expressions for MCD. Thereby the calculated
results for the B-terms are improved considerably. The values for
transition energies, which were quite satisfactory, are hardly affected
whereas the transition moments generally are improved for the alternant
pairing forbidden transitions. Unfortunately, the new set of penetration
integral values does not seem to work for non-benzenoid hydrocarbons,
like biphenylene (147). This may be seen as a result of the different
importance of non-neighbor interactions in the smaller ring.

A very inexpensive, but not necessarily consistent method for
the calculation of MCD B-terms was also suggested (147). This method
is based on PPP wavefunctions with the non-neighbor terms of H^{core}
included only for the calculation of matrix elements of $\vec{M}$. These terms
are not taken into account in connection with the determination of the
wavefunctions, where they without inclusion of other similar terms are
likely to cause a too strong breakdown of the alternant pairing symmetry.
The procedure has some theoretical justification, since the next-neighbor
terms of H^{core}(β_{13}) are an order of magnitude smaller (with opposite
sign) than the matrix elements between neighbors (β_{12}); that is:
$\beta_{13} \sim -0.1\beta_{12}$ (150). The non-neighbor terms are therefore most important
when the contributions from β_{12} vanish, as they do for the calculation
of matrix elements of $\vec{M}$ in a case of perfect pairing symmetry. Additional

Table 3. Alternant hydrocarbons

	E	f	POL.	B	E	f	POL.	B	E	f	POL.	B	E	f	POL.	B
L_b EXP.	31.4	0.002	z	-0.08	29.0	-	(z)	-0.05	29.0	0.004	Y	+0.05	27.0	0.003	Y	+0.2
L_b CALC. 1'	32.8	0.00006	z	-0.15	29.2	0.00001	z	+0.1	29.9	0.001	Y	+0.6	28.2	0.0001	Y	-3.1
L_b CALC. 1	31.9	0.06	z	-4.5	28.7	0.2	z	+23.1	30.3	0.009	Y	-1.6	27.4	0.03	Y	-25.8
L_b CALC. 2	32.4	0	(z)	0	29.0	0	(z)	0	29.9	0	(y)	0	27.9	0	(y)	0
L_a EXP.	35.0	0.3	Y	+2.2	26.9	0.1	Y	+0.9	34.1	0.1	z	+0.2	29.8	0.5	z	+1
L_a CALC. 1'	35.8	0.2	Y	+0.9	27.1	0.3	Y	+0.7	34.0	0.3	z	-0.5	28.9	0.8	z	+4.2
L_a CALC. 1	34.9	0.2	Y	+6.6	27.0	0.3	Y	-19.8	34.2	0.4	z	+3.0	28.6	0.8	z	+29.6
L_a CALC. 2	35.1	0.1	Y	+0.3	27.2	0.3	Y	+0.3	33.5	0.4	z	+0.3	28.4	0.8	z	+0.5
B_b EXP.	45.2	1.6	z	+4	39.5	1.6	z	+2.9	38.8	0.1	Y	+4	36.5	0.4	Y	+3.5
B_b CALC. 1'	46.9	2.2	z	+2.7	40.7	3.0	z	+0.5	39.9	0.4	Y	-7.2	38.3	0.9	Y	+1.8
B_b CALC. 1	46.5	2.0	z	-0.7	42.3	2.8	z	-2.9	40.7	0.5	Y	+32.5	38.9	0.9	Y	-4.1
B_b CALC. 2	45.1	2.0	z	+2.0	39.9	2.8	z	+0.8	40.4	0.6	Y	+20.4	38.8	1.0	Y	+1.5
B_a EXP.	46	1	Y	<0	45.5	0.2	Y	-0.8	39.7	1	z	-1.3	41.3	1.0	z	-0.8
B_a CALC. 1'	51.3	0.6	Y	-3.5	49.3	0.3	Y	+3.1	40.9	1.6	z	-1.5	44.3	1.6	z	-2.8
B_a CALC. 1	51.3	0.6	Y	-1.0	51.0	0.2	Y	-1.5	41.3	1.3	z	-37.5	44.9	1.6	z	-0.5
B_a CALC. 2	50.5	0.6	Y	-2.3	49.4	0.3	Y	-0.9	40.6	1.5	z	-19.1	44.5	1.5	z	-2.7

Observed values for transition energy (E in 1000 cm^{-1}), oscillator strength (f), transition moment directions (POL) and MCD B-terms (B in $10^{-3} \beta_e D^2/cm^{-1}$) are from (6, 151). The calculated results have been obtained according to the methods calc. 1, calc. 1' and calc. 2 described in the text (147).

support for the procedure is found in the recent theoretical work by
Michl (151, 152). It may be added that an interesting theoretical
discussion of Hamiltonians used for descriptions of the MCD effect is
found in ref. (153).

Michl (151) suggested the value $\beta_{13} = -0.15\beta_{12}$ instead of
$\beta_{13} = -0.1\beta_{12}$. The higher value seems to give slightly better agreement
with observed MCD spectra. Table 3 shows the results of a series of
calculations on alternant hydrocarbons (147). Calculation 1 is based on
explicit inclusion of overlap and theoretical values for the penetration
integrals. In calculation 1' overlap is also included, but the modified
values for penetration integrals, described above, have been used. In
calculation 2, the results are based on PPP wavefunctions but with
$\beta_{13} = -0.15\beta_{12}$ included for the calculation of matrix elements of the
magnetic dipole moment operator. The agreement with experiment is
reasonably good for the two calculations (1' and 2) where approximate
or perfect alternant pairing symmetry for the states is present. The
agreement is clearly unsatisfactory in some cases, when alternant
pairing has been completely removed (calc. 1).

It should be emphasized that the price paid, when the inexpensive
method (calc. 2) is used, is the lack of information about transition
moments and MCD B-terms for transitions between states of the same
pairing symmetry, the alternant pairing symmetry forbidden transitions.
Since such transitions often are dominated by vibronic interactions,
the price may be acceptable.

4. MCD of derivatives of alternant hydrocarbons.

So far the effect of alternant pairing symmetry on the MCD of
alternant hydrocarbons has been considered an evil. It might, after all,
turn out to be a blessing in disguise (152). When the π-electron systems
of alternant hydrocarbons are weakly perturbed by substituents or
replacements, the result is usually a breakdown of the pairing symmetry,
resulting in a strong increase in the intensity of the formerly pairing
symmetry forbidden transitions and an often drastic increase of the MCD
effect, including changes of MCD signs. Moreover, the MCD spectra of
the derivatives are usually strongly dependent on the position or type
of the substituent or replacement. This was demonstrated some years
ago for aza-replacements in the pyrene molecule (31). The results are
shown in Figs. 5-8, and in the following it will be demonstrated how
a simple theoretical model based on Hückel orbitals is able to account
for the observed MCD signs (152). As mentioned before, a more detailed
and complete description based on the free electron model is given by

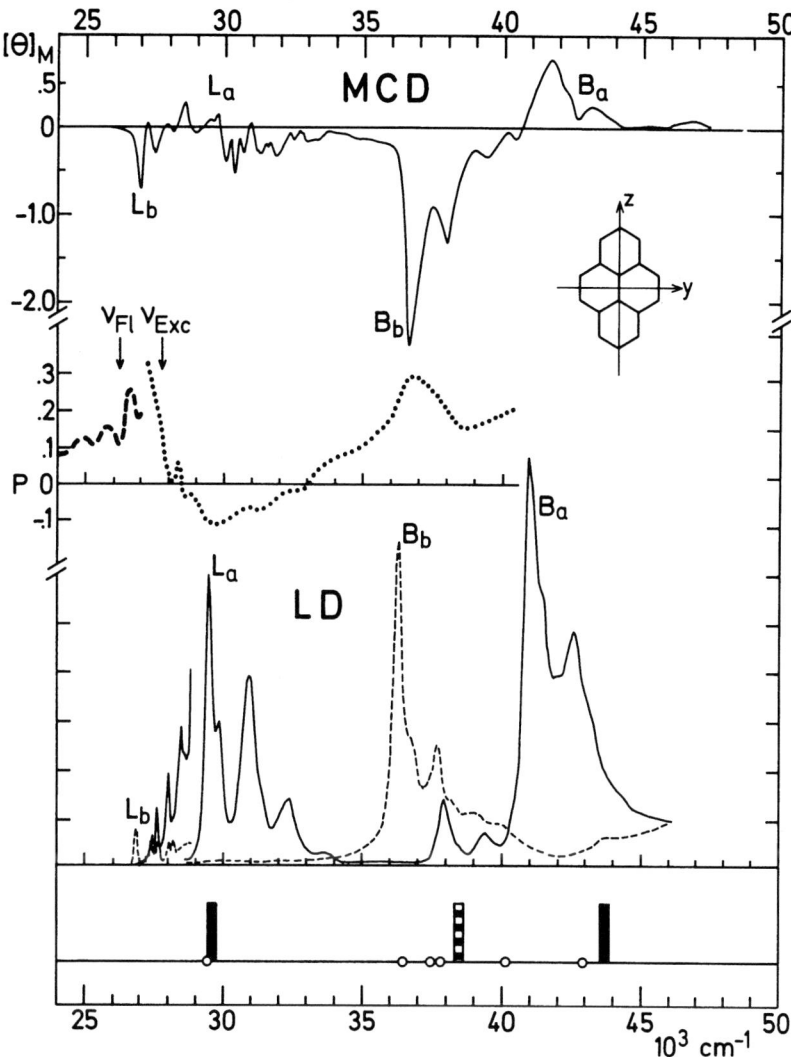

Fig. 5. Pyrene (31). Top: MCD spectrum. Center: degree of polarization for fluorescence and fluorescence excitation spectra. Monitoring wavelengths are indicated. Bottom: z-polarized (full line) and y-polarized (broken line) reduced absorption curves (arbitrary scale). Calculated transitions (PPP) are indicated by full lines (z-polarized), broken lines (y-polarized), and circles (symmetry-forbidden). All calculated B-terms are zero. Oscillator strength is indicated by line thickness.

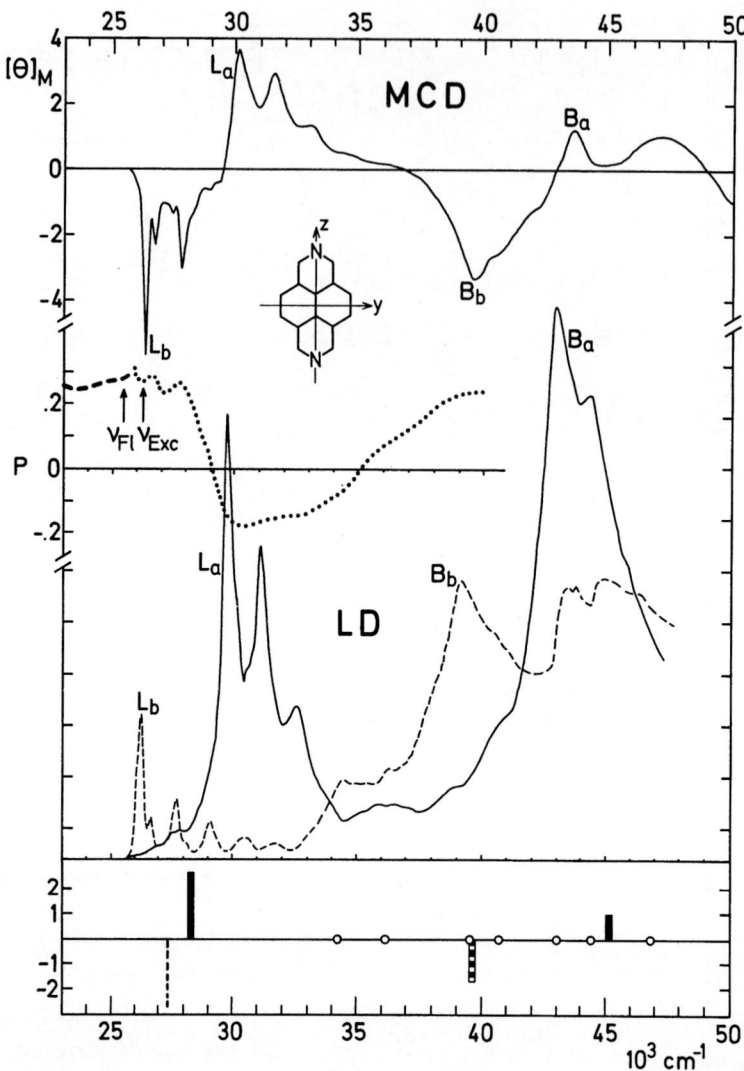

Fig. 6. (31). 2,7-Diazapyrene. See caption to Fig. 5. Length and
direction of the lines indicating calculated transitions give the
value of $-\mathrm{sgn}B[1 + \log|B|]$. The calculated results shown are those
obtained by SCI.

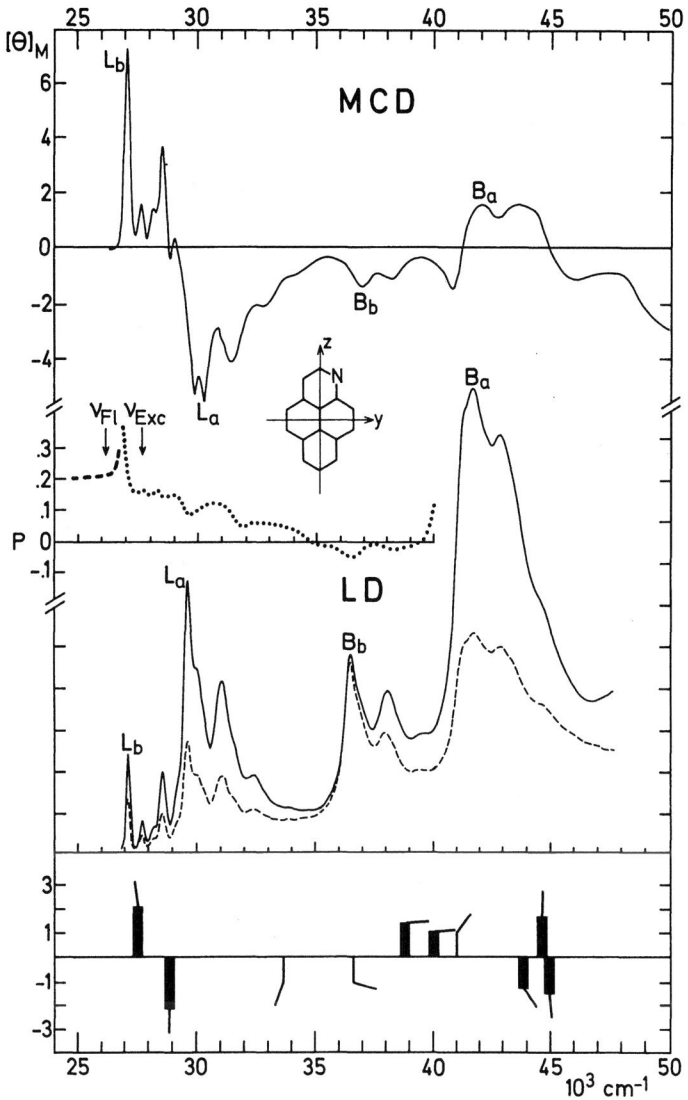

Fig. 7. 1-Azapyrene (31). See caption to Fig. 6. The absorption curves at the bottom are the experimental curves E_Z (full) and E_Y (broken). Calculated polarization directions are shown as directions of flags at the end of lines indicating the transitions.

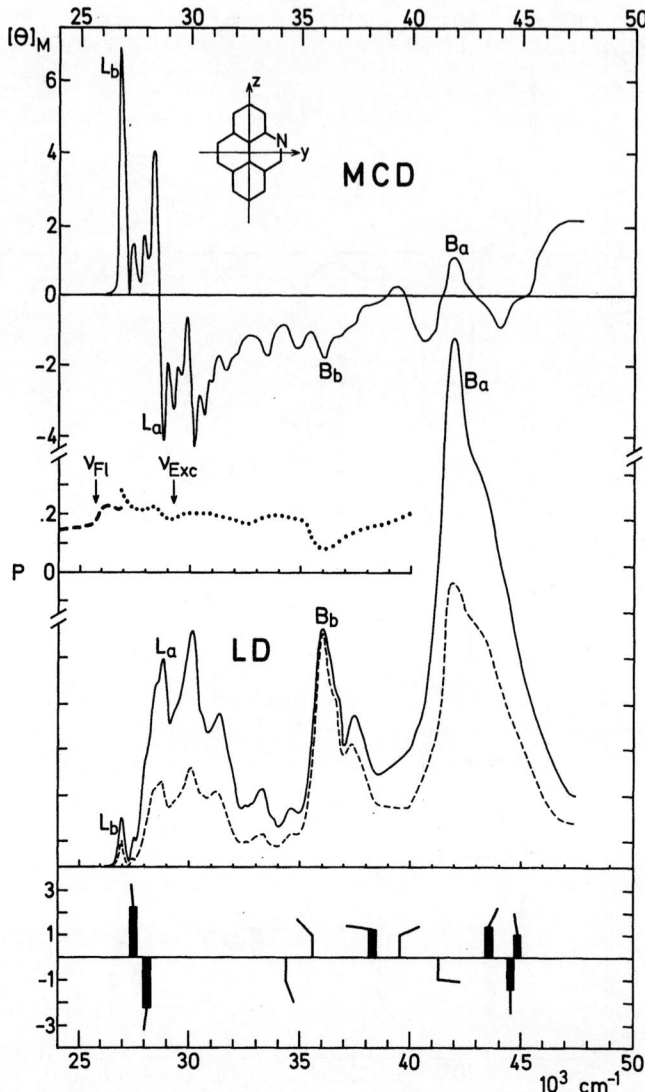

Fig. 8. (31). 4-Azapyrene. See caption to Fig. 7.

Michl (151), and readers interested in the general theory are referred to this work.

According to (IV,2) the mixing in MCD by the first two excited singlet states, usually ^1L_a and ^1L_b, may be written:

$$\langle L_a|-i\vec{M}|L_b\rangle \cdot \langle 0|\vec{M}|L_a\rangle \times \langle L_b|\vec{M}|0\rangle \, (E_{L_b}-E_{L_a})^{-1}$$

The two states ^1L_a and ^1L_b are often very close in energy; thus the B-terms for the transitions from the ground state to the two states may be approximated by:

$$B(^1L_a) = -B(^1L_b) = \langle L_a|-i\vec{M}|L_b\rangle \cdot \langle 0|\vec{M}|L_a\rangle \times \langle L_b|\vec{M}|0\rangle \, (E_{L_b}-E_{L_a})^{-1}$$

$$(IV,3)$$

where it has been assumed that mixing with the ground state and other excited states, which often have considerably higher energies, are of minor importance for the B-terms of ^1L_a and ^1L_b. The two states may be represented by the configurations:

$$L_a = |1 \rightarrow -1\rangle$$

$$L_b = \sin\alpha \cdot |1 \rightarrow -2\rangle - \cos\alpha \cdot |2 \rightarrow -1\rangle$$

where for example $|1 \rightarrow -2\rangle$ describes a configuration obtained from the ground configuration by removing an electron from the highest occupied orbital, 1 (HOMO), and adding an electron in the second lowest unoccupied orbital, -2. The relative contributions from the two configurations are given by the angle α. It can be assumed that the main results of the perturbation are changes of the orbital energies $\{\varepsilon_i\}$ and changes of α from 45°. This is the α-value for a perfectly paired system in which the orbital energies fulfill the relation:

$$\Delta LUMO = \varepsilon_{-2}-\varepsilon_{-1} = \varepsilon_1-\varepsilon_2 = \Delta HOMO$$

From (IV,3) one has (31):

$$B(^1L_a) = -B(^1L_b) = (\sin^2\alpha-\cos^2\alpha)\,K \qquad\qquad (IV,4)$$

where K does not depend on the type of perturbation. The sign of the constant K may be determined from inspection of the orbitals 1, -1, 2 and −2 (152, 31). It is further assumed that non−neighbor matrix elements of the magnetic moment operator are unimportant in the perturbed alternant system and that the sign of $(\sin^2\alpha - \cos^2\alpha)$ in a first approximation may be obtained from the orbital energy differences:

$$\Delta\text{LUMO} = \varepsilon_{-2} - \varepsilon_{-1} \gtreqless \varepsilon_1 - \varepsilon_2 = \Delta\text{HOMO}$$

$$\Longleftrightarrow \varepsilon_{-2} - \varepsilon_1 \gtreqless \varepsilon_{-1} - \varepsilon_2 \Rightarrow E(|2 \to -1\rangle) \lesseqgtr E(|1 \to -2\rangle) \Rightarrow \alpha \lesseqgtr 45^{\circ}$$

In this connection it must be emphasized that especially for the comparison of different molecules, the relation between orbital energy differences and configuration energies is not always straightforward (154).

If the substitution or replacement takes place in position μ, the first order change in the orbital energy ε_i will be:

$$\Delta\varepsilon_{i\mu} = c_{i\mu}^2 \Delta\alpha_{\mu}$$

where $c_{i\mu}$ is the Hückel coefficient for orbital i on atom μ_i and $\Delta\alpha_{\mu}$ is the change in the Coulomb integral at position μ. $\Delta\alpha_{\mu}$ is positive when center μ becomes less electronegative and negative when μ becomes more electronegative. In alternant systems the pairing property ensures that $c_i^2 = c_{-i}^2$. Therefore, the perturbation will lead to $\alpha > 45^{\circ}$ when $\Delta\alpha_{\mu} > 0$ and $c_{1\mu}^2 > c_{2\mu}^2$ or $\Delta\alpha_{\mu} < 0$ and $c_{1\mu}^2 < c_{2\mu}^2$; otherwise, it will lead to $\alpha < 45^{\circ}$.

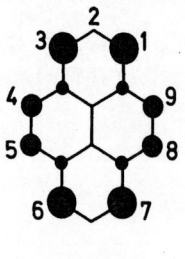

orbital 1

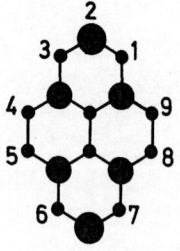

orbital 2

Charge distribution for orbitals 1 and 2 in pyrene.

In the case of pyrene (31) K < 0 and for replacements of C-H
by N, $\Delta\alpha_\mu$ < 0. Therefore an aza replacement in for example position 1
or 4 will lead to α < 45° and B(^{1}L$_a$) = -B(^{1}L$_b$) > 0 whereas an aza-replace-
ment in position 2 will lead to α > 45° and B(^{1}L$_a$) = -B(^{1}L$_b$) < 0. Since
the effect of two or more substituents or replacements is additive in
the simple model, the predictions are in perfect agreement with
experiment (Figs. 6-8).[*] The general result of the model applied to the
pyrene molecule is that substituents or replacements leading to $\Delta\alpha_\mu$ < 0
will give the MCD sign sequence: +,- for the first two transitions, when
μ = 1, 3, 4, 5, 6, 8, 9 or 10, and: -,+ when μ represents the remaining
positions. The opposite sign sequences are obtained when $\Delta\alpha_\mu$ > 0.

In spite of the very approximate nature of the model (152)
such sign predictions usually agree well with experiment, which clearly
makes MCD a potentially very powerful analytical tool. A discussion of
the structural reasons for the agreement can be found in the earlier
mentioned general analysis in the perimeter model of the relation
between MCD signs and other properties of the electronic systems (151).

5. Summary.

A comparison of experimental results for a series of planar
organic molecules with calculations in the π-electron approximation of
transition energies, oscillator stregnths, transition moment directions,
MCD B-terms, and in some cases substitutent effects has been performed
for numerous molecules (3, 4, 6, 8, 22, 24, 29, 31, 43, 55, 62, 64, 147,
155-62). From these comparisons follow that with few exceptions it is
possible to assign several low-lying spin-allowed observed transitions
in each spectrum with reasonable confidence. In some cases transitions
of σ-π*, in particular n-π*, type must be included with the calculated
π-π* transitions. The agreement is usually very good for transition
energies. For oscillator strengths the agreement is good for strong
transitions, but is much less satisfactory for weaker ones. The same
has been observed for transition moment directions of low symmetry
molecules. For molecules with one or two symmetry planes perpendicular
to the molecular plane the moment direction is used as a strong assignment criterion corresponding to a requirement of perfect agreement.

The calculated MCD B-terms agree surprisingly well with the
observations in regions dominated by only a few allowed electronic
transitions. This is even true for many transitions in alternant
hydrocarbons, when the models discussed above are used. In spectral
regions with a large number of close-lying allowed electronic transitions, the calculated B-terms will often deviate in both size and sign

[*] MCD sign = - sign of B-term

from the observations.

The natural next step in the development of theoretical methods for the description of the electronic spectra of planar organic molecules seems to be a more explicit description of the interactions between the σ- and π-electron fields. This may be done by semiempirical all valence electron models (146), or for small to medium-sized molecules maybe even by ab initio techniques. It is encouraging to note that electronic excitation energies for diatomic molecules may be calculated with a reasonable accuracy by inexpensive ab initio methods (163) and recent progress in computational methods (164) and computer technology may make calculations on much larger molecules possible. It should be added, however, that for many organic molecules, e.g. the important group of large molecules with direct biological interest, semiempirical methods may still for many years provide the best possibilities for an economical and reliable description.

One interesting possibility for an experimental investigation of the interaction between the σ- and π-electron fields and of the mixing in MCD by states of π-π* and σ-π* (in particular n-π*) type is provided by studies of MCD of oriented samples (11). Such experiments might be of considerable importance for the development of all valence electron theoretical models.

V. EXAMPLES OF LD AND MCD SPECTRA AND SPECTRAL ASSIGNMENTS.

The methods described above have been illustrated by only a
few real spectra and spectral evaluations. In the following some
characteristic examples of assignments of electronic transitions will
be given in order to further illustrate the procedures recommended.

1. Fluoránthene.

In the widely used book by Birks (165) the UV spectrum of fluoranthene in the energy region below 48000 cm^{-1} is accounted for by only
four excited singlet states. A few years earlier the spectrum was inter-
preted from the assumption of only three transitions in the region
(166, 167).

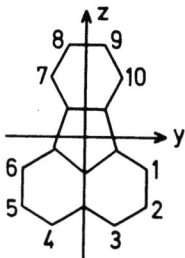

The existence of four transitions is in apparent agreement with the shape
of a solution absorption spectrum of reasonably high quality, although
it has been a problem that the first transition (^1L_b) is almost hidden
below the second band (^1L_a). In spite of the apparent agreement,
standard PPP calculations (168) predict the existence of at least 8
transitions in this energy region. Polarized spectroscopy experiments
such as fluorescence polarization (168) and stretched sheet absorption
spectroscopy (19, 61) proved already in the 1960'ies the existence of
3-4 additional bands. In particular, the stretched sheet data were very
useful in this connection, and Fig. 9 shows some early stretched sheet
results for transitions 2-9 in fluoranthene (19, 61). The short axis
polarized first transition which was too weak to be studied in detail
in the early stretched sheet spectra has later been observed in sheets
(169) at low temperature as well as in fluorescence (168, 170).

Further investigations by means of π-electron calculations
including higher order CI and experimental work including low tempera-

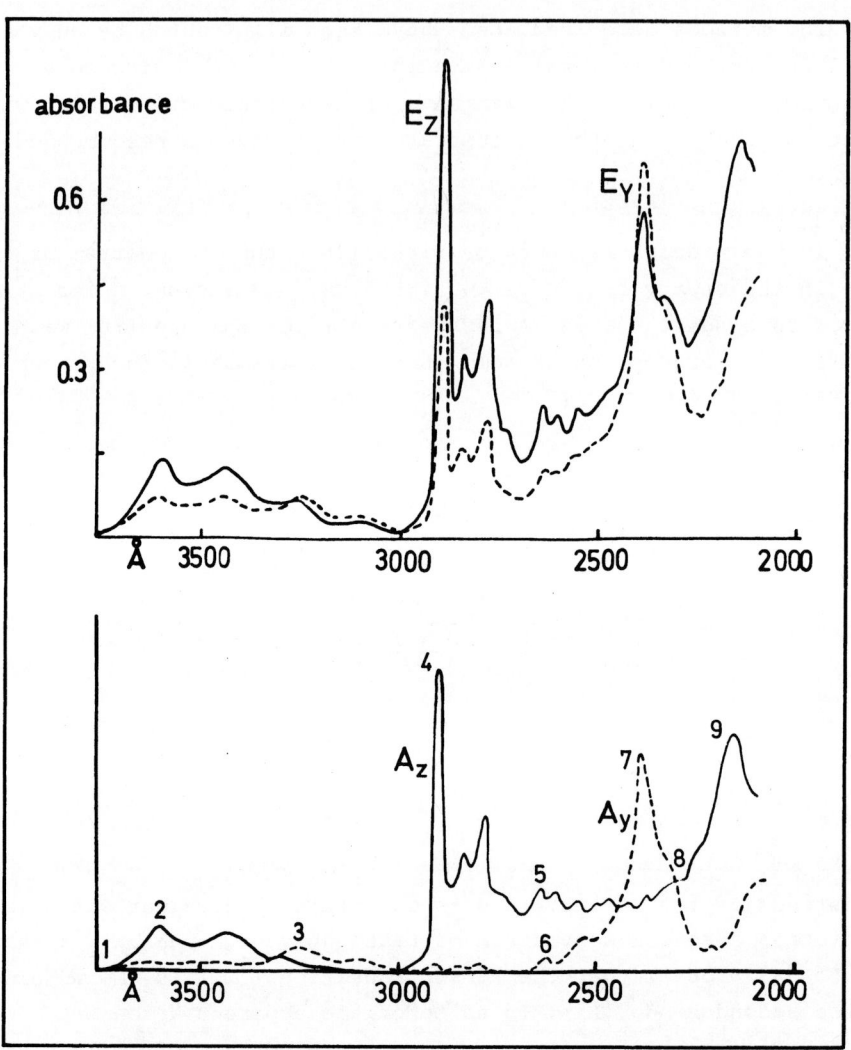

Fig. 9. Fluoranthene (61). Top: observed spectra obtained at room temperature in stretched polyethylene sheets. Bottom: reduced spectra.

ture spectroscopy, fluorescence polarization, additional stretched
sheet experiments, and MCD have later been carried out on both fluoranthene and amino and fluoro substituted fluoranthenes (43, 126),
see also ref. (171). This extensive effort has resulted in the assignment of nine electronic transitions below 48000 cm^{-1} (Table 4).

The existence of the first transition, just below 25000 cm^{-1}
in energy, has been questioned by some authors (172) but it seems now
well established. It has also been claimed that the first excited state
is nonplanar (173), but the evidence does not seem very convincing (43).
The intensity of the first transition is changed by a factor 0.5-2
when a hydrogen atom in the molecule is replaced by fluorine, and the
transition moment direction of this very weak transition changes
considerably (43). Since a mono-substitution leads to a lowering of the
C_{2v} symmetry, the π-π* transitions are no longer limited to the y- and
z-directions. As expected, amino-substitution which corresponds to a
larger perturbation of the π-electron system increases the intensity
considerably; the transition moment directions change so much that they
in some cases (3-, 7- and 8-NH_2-fluoranthene) become almost parallel
to the z-axis (Table 5). The Franck-Condon forbidden shape observed in
fluoranthene and the fluoro-derivatives, with large contributions from
z-polarized components, due to borrowed intensity from the second
transition, is not found in the amino-derivatives (126).

The MCD of the first transition in the parent hydrocarbon is
weak and positive (negative B-term), and is changed relatively little
by F- and even NH_2-substitution. The general agreement between observed
and calculated MCD results is very good, with the exception of the
fairly small substituent effects, which as expected often are calculated
with large relative errors (43, 126). From the combined evidence, however,
the assignment of the first transition cannot be questioned.

The second transition is long-axis (z) polarized, has medium
intensity and a negative B-term. The sign of the MCD is not changed
by F- or NH_2-substitution, but the absorption intensity is decreased
considerably. The assignment of the transition is obvious (Table 4)
and has never been questioned.

The third transition, which is of medium intensity, was overlooked for many years, but is easily seen in the reduced spectrum $A_y(\lambda)$
obtained from the LD absorption spectra (Fig. 9). The MCD for the 0-0
peak is clearly negative (postive B-term) both in the parent hydrocarbon
and in the fluoro-derivatives, but the following vibronic peaks have
all positive MCD, except in the case of 7-F-fluoranthene. The sign
reversal may be contributed to the second transition which has some

Table 4. Fluoranthene

	$E_{obs.}$	$E_{calc.}$	$f_{obs.}$	$f_{calc.}$	$Pol_{obs.}$	$Pol_{calc.}$	B_{obs}	$B_{calc.}$
1	24.8	26.7	0.012	0.01	y	y	-0.2	-1.9
2	27.8	28.1	0.17	0.49	z	z	-2.7	-6.3
3	31.0	31.0	0.05	0.07	y	y	+0.6	+6.6
4	34.7	35.9	0.55	0.22	z	z	+5.3	+0.8
5	37.9	40.6	~0.15	0.06	z	z	-	-
6	38.2	41.0	~0.15	0.05	y	y	-	-
7	42.2	43.2	0.44	1.22	y	y	-	-
8	43.5	42.0	0.15	0.53	z	z	-	-
9	46.9	47.1	0.26	0.81	z	z	-	-
10	-	46.5	-	0.0004	-	y	-	-
11	-	49.9	-	0.04	-	z	-	-
12	~49	50.1	<0.1	0.08	y	y	-	-

Observed values for transition energy (E in 1000 cm^{-1}) and oscillator
strength (f) are from measurements in 3-MP at 77°K. Observed transition
moment directions (Pol) wer determined from measurements in stretched
polyethylene sheets also at 77°K, and the MCD B-terms (B in $10^{-3}\beta_e D^2/cm^{-1}$)
were determined in cyclohexane at room temperature (43). The calculated
values are from PPP calculations including all singly excited configurations with the Mataga-Nishimoto expression for the two-electron repulsion
integrals (43). The MCD B-terms are only given for the first four
transitions, since the higher region has a very high density of states,
which makes a reliable determination of the B-terms difficult.

Table 5

Transition Moment Directions in Aminofluoranthenes (126)

		1 Exp.	1 Calc.	2 Exp.	2 Calc.	3 Exp.	3 Calc.	4 Exp.	4 Calc.	5 Exp.	5 Calc.	6 Exp.	6 Calc.	7 Exp.	7 Calc.
Fl	E	24.7	26.8	27.7	28.3	30.9	31.7	34.7	36.2	37.9	40.9	38.1	41.3	42.0	43.4
	ϕ	90	90	0	0	90	90	0	0	0	0	90	90	90	90
1-NH$_2$	E	24.4	25.5		28.1	30.3	30.4	34.9	35.7		38.0	38.4	39.7	38.5	40.3
	ϕ	-55±5	-40		+15	+50±5	+40	-40±5	0		+70	[+30±10]+30		+65±10 +75	
2-NH$_2$	E	23.0	24.0	27.0	27.8	29.8	30.0	33.6	35.2	38.0	38.5		39.3	40.3	40.5
	ϕ	+52±5	+30	+40±5	+15	-58±8	-50	-30±10	+45	-15±15	-20		-30	(-)68±8	-60
3-NH$_2$	E	~22	24.6	27.0	28.9	30.4	31.0	32.6	35.2	36.2	37.6	39.8	38.2	40.4	41.8
	ϕ	-25±10	-20	+12±12 +10		-12±12 + 5		-12±12	-15	+70±20 +25		(-)22±22 +80		+70±20	-20
7-NH$_2$	E	23.1	22.9	(25.6 – 27.0)	27.1	30.6	31.3	32.7	33.6	35.8	38.0	37.2	38.6	40.2	39.3
	ϕ	+30±10	0	+30±3	+15	-68±8	-100	+30±10	0	+10±10	+30	-30±10	+50	-68±8	-75
8-NH$_2$	E	~22	24.6	27.4	28.3	30.7	31.3	32.6	33.5	36.1	37.4	40.5	38.7	41.7	41.2
	ϕ	+30±10	+30	-10±10	0	-68±8	-85	+10±10	-10	-37±15	+10	-60±20	+85	80-90	+80

Angles ϕ between the effective orientation axis of the molecules (approximately the C$_2$ axis of the fluoranthene skeleton) and the transition moment direction (in degrees). The sign for the angle of the first transition has been chosen so as to agree with that calculated (positive, clockwise; negative, counter-clockwise, with the substituent on the right-hand side of the molecule). Calculated values refer to angles between the C$_2$ axis and the transition moment direction (degrees). Energies (E) are given in units of 1000 cm^{-1}.

vibronic components in the regions 32-34000 cm^{-1}. It may also be explained (43) as a result of different mixing of the 0-0 transition and the following vibronic peaks with transitions 2 and 4: mixing with the former gives negative contributions to MCD, mixing with the latter positive contributions, and these are likely to dominate at higher energy because of the reduced energy difference. The assignment of the third transition is from this evidence also clear.

The fourth transition is strong and shows a detailed vibronic structure. It is z-polarized and the MCD is negative for both fluoranthene and the derivatives. The intensity of the absorption is in some cases drastically reduced by substitution. The assignment is obvious (Table 4, Fig. 9).

Transition 5 is of medium intensity, it is z-polarized with a weak negative MCD. The band corresponding to transition 5 seems to extend to quite high energies. On the basis of substituent shifts observed in both absorption and MCD, the assignment of the band to a separate electronic transition is relatively safe (43).

The existence of the sixth transition, which is y-polarized and of medium intensity, was first proposed on the basis of stretched sheet spectra (61). Because of the Franck-Condon forbidden shape it was not quite clear, however, whether the band was due to a separate electronic transition or should be assigned to vibronic components of transition or should be assigned to vibronic components of transition 5 borrowing intensity from the strong y-polarized transition 7. This question is easily solved by a study of substituent shifts for the fluoroderivatives both in absorption and particularly in MCD (43). Therefore, the assignment of this transition seems now well established (Table 4, Fig. 9).

The seventh transition is strong, y-polarized and has positive MCD (negative B) in all compounds except 2-NH$_2$-fluoranthene. The assignment seems quite obvious. As mentioned in chaper III a mixed polarization of this transition has been proposed in an LD stretched sheet investigation of fluoranthene (70) due to the use of an incorrect reduction procedure (the unrealistic assumption of a rod-like orientation distribution, according to the Fraser-Beer model).

The isotropic solution absorption spectrum shows a medium intense shoulder on the high energy side of transition 7. This shoulder has z-polarized components and these may be due to transition 8. Also contributions from one or more transitions different from 7 seem to be present in the MCD in this region. It may, however, still be possible that the absorption assigned to transition 8 is due to vibronic components of transition 7, borrowing intensity from transition

9 through a non-totally symmetric vibrational mode.

The assignment of the z-polarized transition 9 is obvious in spite of the fact that there seems to be little similarity between the shape of the MCD curve and the absorption spectrum. This can undoubtedly be seen as a result of the presence of weaker transitions, which due to small energy denominators in the expression for their B-terms, are likely to appear with much higher relative intensity in the MCD spectrum than in the absorption spectrum.

2. Thiophene.

While the usefulness of MCD in connection with the assignment of the electronic transitions in fluoranthene was limited to some of the transitions, the LD results were of general importance. In the case of a symmetrical planar molecule of near perfect disc-shape, the resulting orientation distribution in a stretched sheet will fulfill the relation $K_z \sim K_y$. Thus, the information obtained from stretched sheet spectroscopy is of little use in connection with the assignment of the inplane polarized transitions since according to (III,6) $A_y(\lambda)$ and $A_z(\lambda)$ cannot be separated. An example of this orientation distribution is provided by the thiophene molecule (4).

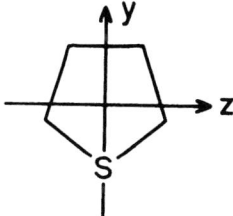

The first absorption band of thiophene (Fig. 10) has been discussed by several authors (174, 175) and there has been some disagreement with respect to the number of transitions responsible for the band.

The MCD spectrum (Fig. 10) shows clearly that the absorption band is composed of two transitions, the first with positive, the second with negative MCD. Calculations in the PPP model predict the presence of two almost degenerate transitions, one z- and one y-polarized, the lower with positive, the higher with negative MCD. Since the two B-terms are almost completely due to mixing between the two

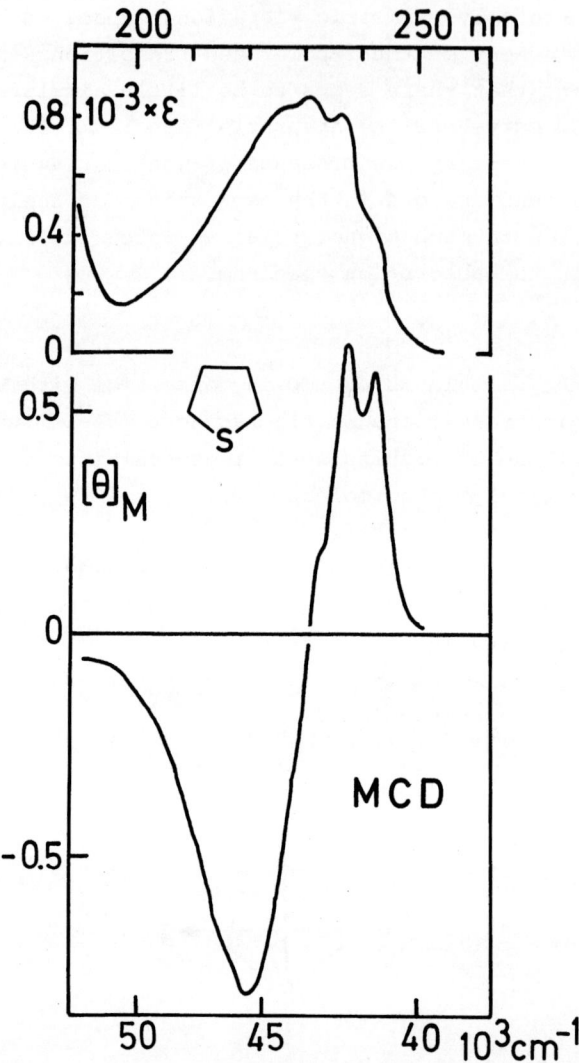

Fig. 10. Thiophene (4). Top: absorption spectrum in n-hexane at room temperature. Bottom: MCD spectrum in units of deg. m^{-1} ℓ mol^{-1} $gauss^{-1}$ obtained at room temperature.

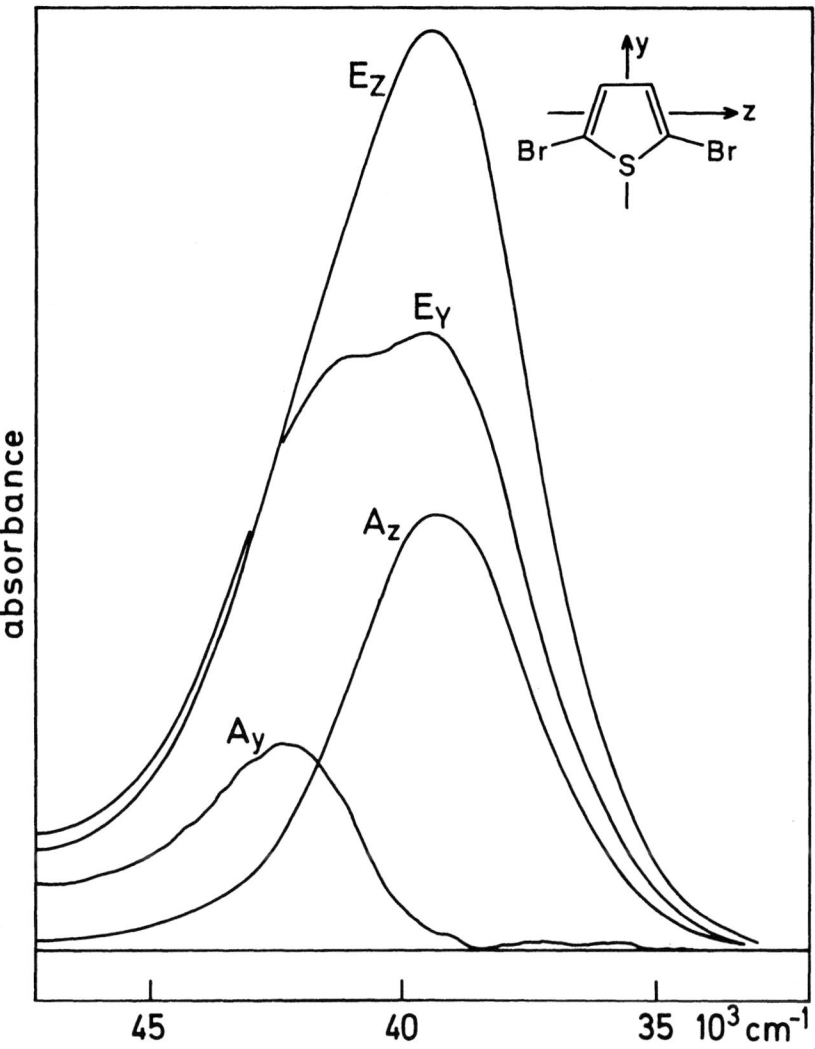

Fig. 11. 2,5-Dibromothiophene (161). Observed spectra obtained at room temperature in stretched polyethylene sheets, and reduced spectra.

transitions, a change in their calculated order will also change the
calculated MCD signs. This may be seen as a result of a sign change of
the energy difference denominator in (IV,2). Then, the lower of the states
will still be predicted to have positive MCD. Therefore, the signs
observed in the MCD spectrum cannot be used for a determination of the
energy order of the two transitions, although it revealed their
presence.

Studies of 2,5-dibromothiophene (161), which has an MCD spectrum
almost identical to that of thiophene and aligns in a stretched sheet
such that $K_z > K_y$, show that the absorption is composed of a long wavelength z-polarized part and a weaker y-polarized component at higher
energies (Fig. 11). This further confirms the presence of two transitions in the energy region, but since the substituent shifts are
not exactly known, the seemingly obvious assignment of the first
(postitive) MCD peak in thiophene to a z-polarized, and the second
(negative) to a y-polarized transition still cannot be claimed to be
fully established. It should be noted that the real energy difference
between the two transitions may be strongly overestimated by the
energy difference between the two MCD peaks seen in Fig. 10. The
reason is compensation effects due to overlap between the positive and
negative contributions to the MCD. These effects tend to shift the
peaks away from each other, and the two transitions may be extremely
close in energy.

MCD spectra (161) of two compounds related to thiophene, name-
ly selenophene and tellurophene, show strong similarities to the
spectrum shown in Fig. 10. Also MCD spectra of a series of thiophene
derivatives are very similar to the spectrum of thiophene (161). In
other words, thiophene is a "hard" MCD chromophore. MCD spectra of
pyrrole, furan, and some derivatives show rather unexpectedly that
these molecules are "soft" chromophores, and the complete assignments
remain a problem (161), even after inclusion of LD studies of
derivatives of the two molecules.

3. Acenaphthylene.

When modern applications of the stretched sheet technique were being attempted in the early 1960'ies (6, 19), one major concern was the applicability of the method. Naturally, one of several questions asked was: can only very elongated molecules be oriented sufficiently in stretched polymer sheets, or is the technique applicable to organic molecules with almost any shape? After the initial development of the technique (choice of polymer, solvent, degree of stretching, etc.) acenaphthylene was chosen as the test molecules for the orienting power of polyethylene sheets.

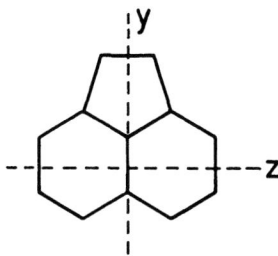

Since the difference in length between the two in-plane dimensions of the molecule is very small, it might be expected that the orientation would correspond to that of a disk ($K_y = K_z$, the left side of the orientation triangle in Fig. 2). However, the spectra obtained showed a significant difference in the alignment of the molecular y- and z-axes (6,19). At low temperature (64) the orientation corresponds to $(K_z, K_y) = (0.43, 0.38)$. However, as mentioned in chapter III, the y- and z-axes are interchanged and the values $(K_z, K_y) = (0.54, 0.33)$ are obtained when the hydrogen atoms in positions 1 and 2 in acenaphthylene are replaced by bromine atoms. The orientation effect in the sheets is in general very sensitive to small changes of the structure of the solute molecules (Fig. 2).

The stretched sheet spectra of acenaphthylene and the bromine derivative do not allow an immediate assignment of the transitions. For example is the dichroic ratio $E_z(\lambda)/E_y(\lambda)$ for both molecules larger than or equal to 1 in the whole spectral range, including both y- and z-polarized transitions. This is in agreement with the fact that for both molecules $K_y \geq 1/3$. The value of the experiment is greatly enhanced when reduced spectra are constructed. These are shown in Figs. 12 and 13, and allow identification of at least 5 transitions below 45000 cm^{-1}. Comparison with various π-electron calculations of the PPP-type including calculations which involve a large number of doubly excited configurations (64) has resulted in assignments of

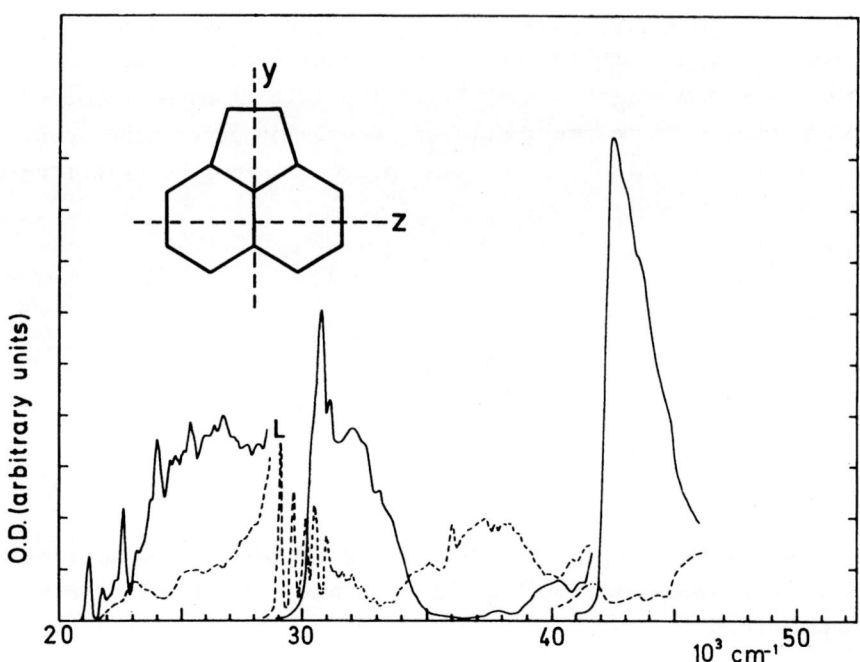

Fig. 12. Acenaphthylene (64). Reduced spectra $A_z(\lambda)$ (full line) and $A_y(\lambda)$ (dotted line), obtained from stretched sheet spectra recorded at ~10 K by means of a closed-cycle helium refrigerator. For the low-energy part of the spectrum a "thick sheet" made from a large number of ordinary sheets (see chapter II) was used in order to ensure a sufficient optical density.

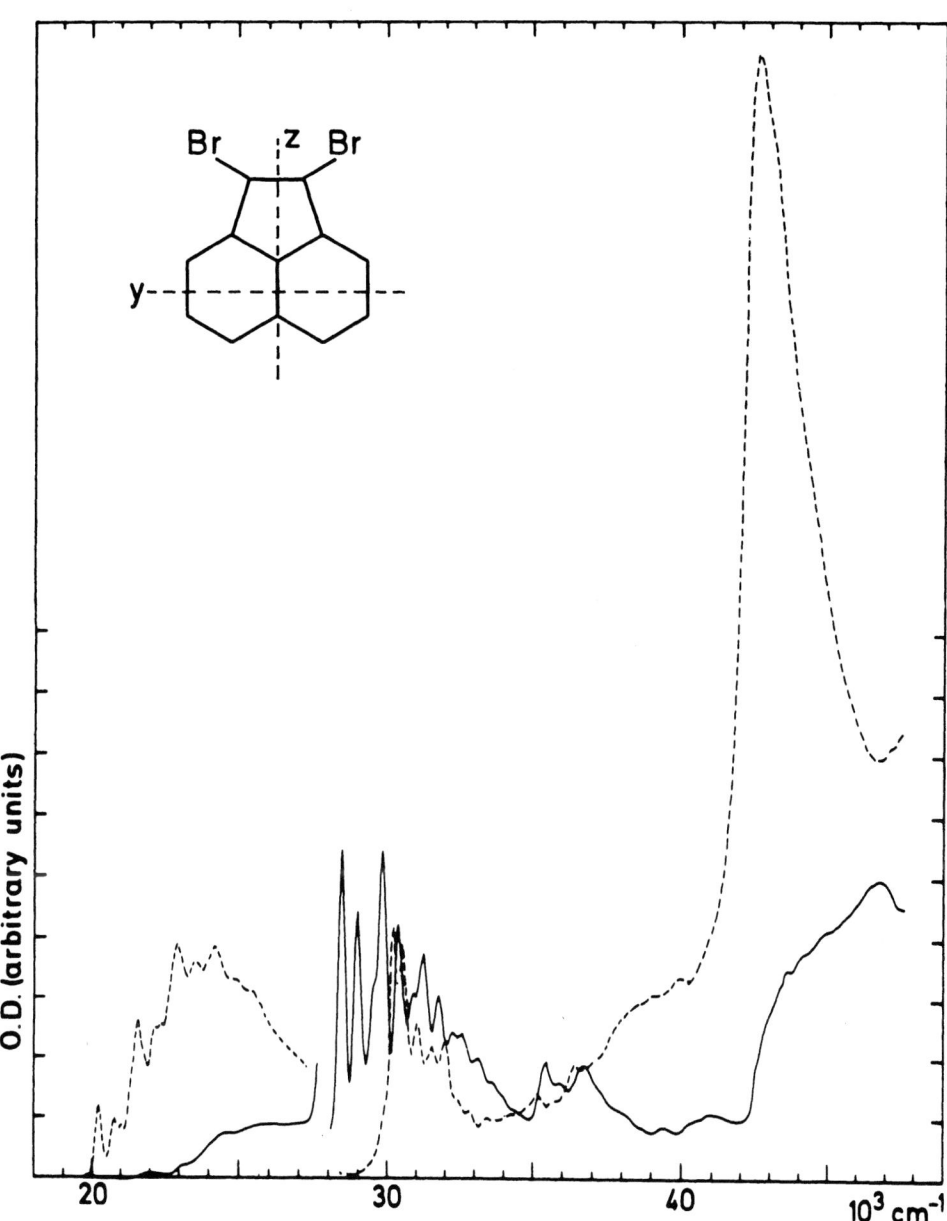

Fig. 13. 1,2-Dibromoacenaphthylene (64). See caption to Fig. 12.

these 5 transitions. MCD spectra of the two compounds confirm this assignment and indicate the presence of a sixth transition with a positive B-term (64) near 50000 cm^{-1}. Contrary to the first 5 transitions this transition is not well described by calculations involving only singly excited configurations. Therefore, an assignment to a calculated transition to an excited state which is predominantly of doubly excited nature has been proposed (64).

4. p-Terphenyl

While acenaphthylene is oriented relatively poorly in a stretched polymer sheet, p-terphenyl and its azaderivatives are examples of the best aligned molecules (Fig. 2).

For p-terphenyl the values $(K_z, K_y) = (0.85, 0.08)$ are obtained; this corresponds to a near rod-like orientation $(K_x \sim K_y)$. Azaderivatives such as 3,6-diphenyl-s-tetrazine show almost the same orientation constants (62); this is different (6) from the results obtained on the series anthracene, acridine, and phenazine (Fig. 2).

The reduced spectra of p-terphenyl (62) are presented in Fig. 14 which also shows the results of PPP π-electron calculations. The assignment is relatively straightforward; the most interesting detail is the short-axis (y) polarized intensity hidden under the strong long-axis polarized band near 35000 cm^{-1}. It may be assigned to one or more of the 3 symmetry-forbidden transitions calculated in this region. Two of these are forbidden due to alternant pairing symmetry; in the D_{2h} point group symmetry both of them correspond to y-polarized

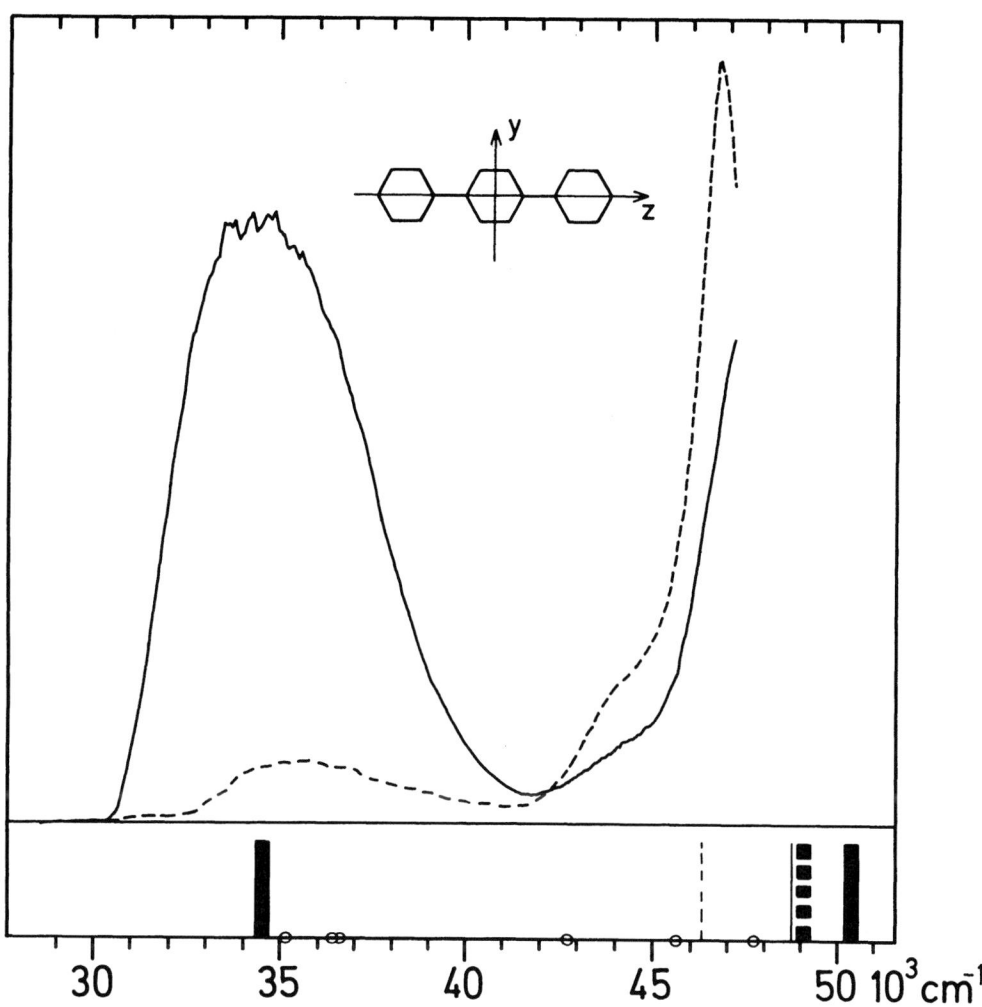

Fig. 14. p-Terphenyl (62). Reduced spectra A$_z$(λ) (full line) and
A$_y$(λ) (dotted line) obtained from stretched sheet spectra recorded at
77 K. Below are shown the results of PPP π-electron calculations. Full
and broken bars correspond to z- and y-polarized transitions respectively, and the calculated oscillator strengths are indicated by the
thickness. Forbidden transitions are indicated by circles.

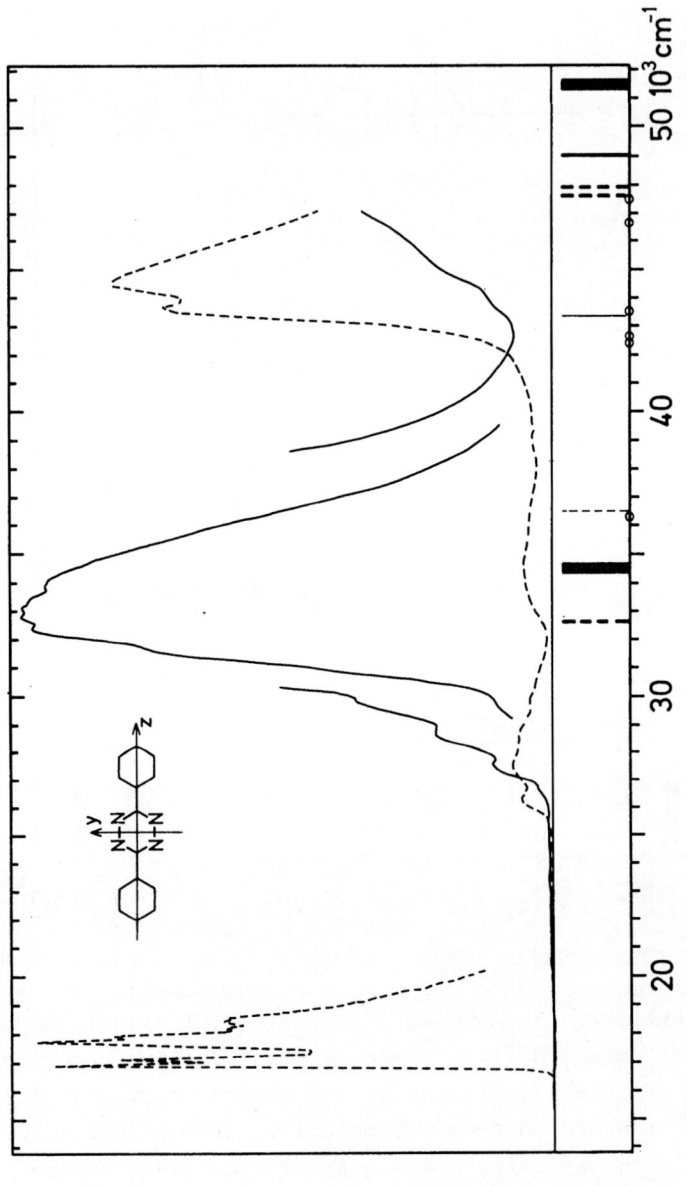

Fig. 15. 3,6-Diphenyl-s-tetrazine (62). Reduced spectra $A_z(\lambda)$ (full line) and $A_y(\lambda) + A_x(\lambda)$ (dotted line) obtained from stretched sheet spectra recorded at 77 K. Below calculated $\pi-\pi^*$ transitions; see caption to Fig. 14.

intensity. It should be added that the assumption of perfect planar
geometry does not hold under all conditions as discussed later.

The spectrum of 3,6-diphenyl-s-tetrazine shown in Fig. 15 is
particularly interesting because of the low energy band near 18000
cm^{-1}. In the planar geometry this can be assigned to a transition
of n-π* type, and should thus be polarized along x. Since π-π* transitions polarized along y and z also are present, the molecule offers
a possibility for an independent determination of both K_x, K_y, and
K_z. The result is $(K_x, K_y, K_z) = (0.075, 0.08, 0.85)$ which confirms the
predicted near rod-like behaviour and agrees with the relation
$K_x+K_y+K_z = 1$.

Fig. 15 also shows that the 2 transitions near 35000 cm^{-1},which
were alternant pairing symmetry forbidden in p-terphenyl, have become
allowed and that one of them has moved to lower energies, below the
strong, allowed z-polarized band. The assignment of at least 7 transitions is possible according to Fig. 15 (62); the double peak near
44000 cm^{-1} seems to correspond to two y-polarized electronic transitions.

The geometry of p-terphenyl, in particular its planarity,
has been investigated by means of x-ray and neutron diffraction methods.
The results are that the molecule in crystals is twisted at low temperature but almost planar at room temperature (176-178). X-ray diffraction
studies of 3,6-diphenyl-s-tetrazine show that this molecule is almost
planar (179). In the gas phase or in solution the angle between the
rings in p-terphenyl seems to be less than or equal to 20° (180-182).
In spite of such variations in molecular geometry, the UV absorption
spectra of p-terphenyl in gas phase, solution and solid phase show no
unusual differences (183-185). This is consistent with the results
shown in Fig. 14 and with the satisfactory assignment to transitions
calculated in the π-electron approximation based on the planar geometry.

VI. CONCLUSIONS

Some important conclusions may be drawn from the results described in the preceding chapters. They deal with the approach to the
assignment process, the use of orientation models for interpretation
of LD spectra, the theoretical description of the lower electronic
transitions in planar organic molecules, and with the possible new
applications of information obtained by MCD spectroscopy.

1. The assignment process.

The assignment of electronic transitions in molecules is a
process based on assignment criteria. The quality of the theoretical
model and the experimental results available are determining factors
for the efficiency of these criteria. It is in general important that
all available criteria are used. Many examples of incomplete or even
incorrect assignments found in the literature can be ascribed to the
fact that the authors used too few and too ineffective criteria. Two
criteria that have demonstrated their special effectiveness for the
electronic transitions in organic molecules are the transition moment
directions and the MCD B-terms. In the case of the UV spectrum of the
fluoranthene molecule these two additional criteria made it possible
to assign 9 electronic transitions in a region where formerly only 4
transitions were found.

Both LD and MCD spectroscopy are today experimentally quite
simple, and the interpretation of the LD and MCD spectra should not
in general cause serious problems. In view of the often very essential
information obtained by these methods they deserve more frequent use.

2. Interpretation of LD spectra.

The most efficient experimental methods in LD spectroscopy deal
with permanent or photo-selected partially oriented samples. Examples
are fluorescence polarization and stretched sheet or liquid crystal
methods. In the fluorescence polarization technique, the orientation
distribution function is known. In the methods based on anisotropic
solvents like stretched polymer sheets or liquid crystals, the orientation distribution is usually unknown. The possibilities for a complete
determination, for example by means of x-ray diffraction, seem in these
cases rather small. Many authors have found it tempting to make
assumptions of specific orientation distribution functions, which
cannot be proven experimentally, but which in many cases account for
the experimental LD data available. Such experimental LD results are

based on one-photon processes like absorption and correspond in general
to an infinity of possible orientation distributions with common values
of (K_x, K_y, K_z). It seems therefore very unlikely that the specific
assumptions will work for new types of experimental information
involving averages of higher powers of the directional cosines, such
as two-photon processes (polarized Raman, fluorescence or two-photon
absorption).

In some cases the conclusions based on experimental LD spectra
are clearly incorrect because of limitations in the model used and
special assumptions about the orientation distribution function. This
is particularly unfortunate, since the experimental basis for the
interpretation often has been of high quality.

The treatment of LD spectra first proposed by Eggers, Michl,
and Thulstrup (19, 24, 61) is not based on assumptions about the
orientation distribution function, and involve only orientational
parameters that can be determined experimentally. The method still seems
to be the best available for dealing with unknown orientation
distributions. The orientation triangle shown in Fig. 2 (7, 54, 64)
will in many cases make an estimate of the necessary orientation
constants possible.

It must be emphasized that it is extremely important that all
information available is extracted from the experimental spectra. A
single curve describing the variation of the dichroic ratio through
the spectrum is not in general sufficient for a satisfactory understanding, and conclusions based on such incomplete data have often
been incorrect.

3. Theoretical descriptions of electronic transitions.

The attempts to describe observed LD and MCD spectra of planar
organic molecules in the energy region below 50000 cm^{-1} in the π-
electron approximation have generally been successful. Some important
exceptions which cannot be studied in this model are n-π* transitions,
which are often easy to recognize, Rydberg transitions, usually observed
only in gas-phase spectra at high energies, and possibly other σ-π*
or π-σ* transitions particularly in small organic molecules. Few of
these transitions with energies below 50000 cm^{-1} have caused serious
problems for the assignment in the past.

With the exception of spectral regions with a high density of
transitions, MCD spectra have generally been surprisingly well described
by approximate π-electron models. This indicates that mixing of π-π*
states with states of different types (n-π*, or other σ-π*, π-σ*, or

Rydberg transitions) may be of realtively modest importance. In particular, MCD signs for a large number of molecules have been predicted
correctly by Michl's general description based on the perimeter model
(151). In spite of the success of the PPP model it is likely that the
size of the B-terms in many cases may be changed by mixing with other
states. It is therefore obvious that investigations by means of all-
valence-electron or even ab initio calculational methods of these
problems have considerable interest. Also new experimental information
such as that obtained from MCD spectra of partially oriented samples
(11b) may provide information about the mixing in MCD of π-π* states
with other states.

It is important to note that since the PPP-model is able to
describe the MCD B-terms of the low energy transitions of most π-
electron systems quite well and since no MCD data were available when
the PPP model was proposed, the results obtained lend considerable
support to the philosophy of this semiempirical model.

It seems likely that the PPP-model with the necessary modifica-
tion for alternant systems still for some years will provide the best
possibility for an inexpensive and reasonably reliable description of
the low energy electronic transitions, including the MCD, for planar
organic molecules.

4. New applications of MCD spectroscopy.

A few years ago, a leading research group wrote that the
"interpretation of the MCD spectra of the vast majority of organic
compounds is likely to be empirical in nature" (186). Fortunately,
this statement has not come true. On the contrary, as described above,
calculations of MCD spectra, especially the B-terms, and predictions
of MCD signs are now carried out economically and with reasonable
accuracy at least in spectral regions corresponding to a low density
of electronic states. Simple rules based on Hückel theory, such as
those discussed in Chapter IV for derivatives of alternant hydrocarbons,
are experimentally well fulfilled. Moreover, the general rules (151)
recently given by Michl for MCD signs of a large number of different
types of organic π-electron systems have improved the understanding of
the MCD spectra of organic molecules considerably.

The improved understanding of the MCD spectra, together with
the strong dependence on the type and position of substituents or
replacements which MCD spectra of many molecules exhibit, may be used
in several ways. First, analytical applications of MCD-spectroscopy are
obviously possible, e.g. for the structure determination of derivatives

of both alternant hydrocarbons and other molecules with a weak MCD
of their own. Second, a classification of π-electron chromophores based
on their MCD and of substituents and replacements based on their effect
on the MCD spectra of the parent compounds, may be of great value.

ACKNOWLEDGEMENT

The results presented in this paper would not have been obtained
without the inspiration and cooperation of a large number of fellow
scientists. I am deeply grateful to Jürgen H. Eggers, Jan Linderberg,
and Josef Michl, also for their invaluable friendship and support.
The enjoyable cooperation with Rolf Gleiter and his excellent group,
and with Bengt Nordén is warmly acknowledged. The hospitality and
interest offered by several foreign institutions, in particular the
Quantum Theory Project and the Chemistry Department at the University
of Florida, the Chemistry Department of the University of Utah, and
the Department of Organic Chemistry at Technische Hochschule in
Darmstadt, have been a great encouragement over the years.

The first draft of this paper was typed by my wife Lizzi, and
the final version by Hanne Kirkegaard and Anni Stenstrup. Extensive
technical assistance was kindly given by Arne and Carl Aage Lindahl.
I am grateful to them all.

The present results would not have been possible without the
constant support of the Chemistry Department of the University of
Aarhus. Also several smaller grants, in particular from NATO are
gratefully acknowledged.

REFERENCES

(1)   E.W. Thulstrup, Int. J. Quantum Chem. $\underline{3S}$, 641 (1970).

(2)   J.R. Platt, J. Mol. Spectrosc. $\underline{9}$, 288 (1962).

(3)   E.W. Thulstrup, J. Michl and C. Jutz, Trans. Faraday Soc. $\underline{71}$, 1618 (1975).

(4)   T. Håkansson, B. Nordén and E.W. Thulstrup, Chem. Phys. Lett. $\underline{50}$, 305 (1977).

(5)   A.C. Albrecht and W.T. Simpson, J. Chem. Phys. $\underline{23}$, 1480 (1955).

(6)   E.W. Thulstrup, Thesis, Aarhus University, 1966; Thesis awarded the Aarhus University Gold Medal in Chemistry, 1969.

(7)   E.W. Thulstrup, M. Vala, and J.H. Eggers, Chem. Phys. Lett. $\underline{7}$, 31 (1970).

(8)   J. Michl, E.W. Thulstrup, and J.H. Eggers, Ber. Bunsenges. Physik. Chem. $\underline{78}$, 575 (1974).

(9)   D.J. Caldwell and H. Eyring, "Theory of Optical Activity", Wiley Interscience, N.Y. 1971; P.N. Schatz and A.J. McCaffery, Q. Rev. Chem. Soc. $\underline{23}$, 552 (1969).

(10)   A major contribution was made by O. Halkjær Jensen (1950-79).

(11)   a. J. Linderberg, private communication; b. J. Michl and E.W. Thulstrup, to be published.

(12)   F.S. Richardson and J.P. Riehl, Chem. Rev. $\underline{77}$, 773 (1977); J.P. Riehl and F.S. Richardson, J. Chem. Phys. $\underline{68}$, 4266 (1978); and references therein.

(13)   L. Velluz, M. Legrand and M. Grosjean, "Optical Circular Dichroism", Academic Press, 1965.

(14)   B. Nordén and Å. Davidsson, Acta Chem. Scand. $\underline{26}$, 842 (1972); Å. Davidsson and B. Nordén, Spectrochim. Acta $\underline{32A}$, 717 (1976); Chem. Scripta $\underline{9}$, 49 (1976); B. Nordén, F. Tjerneld, and E. Palm, Biophys. Chem. $\underline{8}$, 1 (1978).

(15) B. Nordén in "Linear Dichroism Spectroscopy", Proceedings of a
      Nobel Workshop, University of Lund (1977) and in "Problems in
      Contemporary Biophysics" Vol. 3 Polish Scientific, 1977;
      B. Nordén, Appl. Spectrosc. Rev. 14, 157 (1978).

(16) H.P. Jensen, Thesis, Technical University of Denmark, 1978;
      H.P. Jensen, J.A. Schellman, and T. Troxell, Appl. Spectrosc. 32,
      192 (1978).

(17) F. Dörr and M. Held, Angew. Chemie 72, 287 (1960); F. Dörr, ibid.
      78, 457 (1966); F. Dörr in "Creation and Detection of the
      Excited State", Vol. IA, M. Dekker, New York, 1971.

(18) H. Ambronn, Ber. Deut. Botan. Ges. 6, 85, 226 (1888), Ann. Phys.
      34, 340 (1888); A. Jablonski, Acta Phys. Polonica 4, 371 (1935);
      P. Pringsheim, ibid. 4, 331 (1935); K.R. Popov, Opt. Spectrosc.
      3, 579 (1957); Y. Tanizaki and N. Ando, J. Chem. Soc. Japan,
      Pure Chem. Sec. 78, 542 (1957).

(19) J.H. Eggers and E.W. Thulstrup, Lectures at the 8th European
      Congress on Molecular Spectroscopy, Copenhagen, 1965.

(20) B. Nordén, Chem. Scripta 7, 167 (1975); J. Konwerska-Hrabowska
      and J.H. Eggers, Spectrosc. Lett. 10, 441 (1977).

(21) Å. Davidsson and B. Nordén, Tetrahedron Lett. 30, 3093 (1972).

(22) J. Spanget-Larsen, R. Gleiter, R. Haider and E.W. Thulstrup,
      Mol. Phys. 34, 1049 (1977).

(23) P.B. Pedersen, J. Michl, and E.W. Thulstrup, manuscript in
      preparation.

(24) E.W. Thulstrup, J. Michl, and J.H. Eggers, J. Phys. Chem. 74,
      3868 (1970).

(25) D. Fournier, A.C. Boccara, and J. Badoz, Appl. Phys. Lett. 32,
      640 (1978).

(26) A.G. Bell, Philos. Mag. 11, 510 (1881).

(27) R.B. Somoano, Angew. Chem. Int. Ed. Engl. 17, 238 (1978).

(28) M. Gisin, J. Michl, and E.W. Thulstrup, to be published.

(29) E.W. Thulstrup, P.L. Case, and J. Michl, Chem. Phys. $\underline{6}$,
      410 (1974).

(30) R. Gleiter, J. Spanget-Larsen, E.W. Thulstrup, I. Murata,
      K. Nakasuji, and C. Jutz, Helv. Chim. Acta $\underline{59}$, 1459 (1976).

(31) E.W. Thulstrup, J.W. Downing, and J. Michl, Chem. Phys. $\underline{23}$,
      307 (1977).

(32) J. Spanget-Larsen and E.W. Thulstrup, unpublished results.

(33) W. Kuhn, H. Dührkopp, and H. Martin, Z. Physik. Chem. $\underline{B45}$,
      121 (1940).

(34) H. Labhart, Chimia $\underline{15}$, 20 (1961); Helv. Chim. Acta $\underline{44}$, 447,
      457 (1961).

(35) W. Liptay and J. Czekalla, Z. Naturforsch. $\underline{15a}$, 1072 (1960);
      W. Liptay, Z. Naturforsch. $\underline{20a}$, 272 (1965).

(36) J.V. Champion, D. Downer, G.H. Meeten, and L.F. Gate, J. Phys.
      E $\underline{10}$, 1137 (1977); J. Breton and G. Paillotin, Biochim. Biophys.
      Acta $\underline{459}$, 58 (1977).

(37) A. Wada, Appl.Spectrosc. Rev. $\underline{6}$, 1 (1972).

(38) A.S. Davydov, "Theory of Molecular Excitons", McGraw-Hill, 1962.

(39) D.S. McClure, J. Chem. Phys. $\underline{22}$, 1668 (1954); ibid. $\underline{24}$,
      1 (1956).

(40) H.C. Wolf in "Advances of Solid State Physics", Vol. $\underline{9}$,
      Academic Press (1959).

(41) E. Sackmann in "Applications of Liquid Crystals", Springer-
      Verlag, 1975.

(42) H. Wedel and W. Haase, Chem. Phys. Lett. $\underline{55}$, 96 (1978).

(43) J. Kolc, E.W. Thulstrup, and J. Michl, J. Am. Chem. Soc. $\underline{96}$,
      7188 (1974).

(44) F. Weigert, Verhandl. Deut. Phys. Ges. $\underline{1}$, 100 (1920).

(45) H. Zimmermann and N. Joop, Ber. Bunsenges. Physik. Chemie $\underline{64}$,
      1215 (1960); $\underline{65}$, 61, 66, 138 (1961).

(46) A.H. Kantalar and A.C. Albrecht, Ber. Bunsenges. Physik. Chemie 68, 361 (1964); A.C. Albrecht and W.T. Simpson, J. Am. Chem. Soc. 77, 4454 (1955); A.C. Albrecht, J. Chem. Phys. 27, 1413 (1957).

(47) J. Czekalla, W. Liptay, and E. Dollefeld, Ber. Bunsenges. Physik. Chemie 68, 80 (1964); W. Liptay in "Modern Quantum Theory, Istanbul Lectures III", Academic Press (1965).

(48) J.J. Dekkers, G. Ph. Hoornweg, C. MacLean, and N.H. Velthorst, Chem. Phys. Lett. 19, 517 (1973); J.J. Dekkers, Thesis, The Free University of Amsterdam, 1979.

(49) G. Aviv, L. Margulies, J. Sagiv, A. Yogev, and Y. Mazur, Spectrosc. Lett. 10, 423 (1977).

(50) J.P. Jarry and L. Monnerie, J. Polym. Sci. Polym. Phys. Ed. 10, 2135 (1972).

(51) E.W. Thulstrup and J. Michl, J. Chem. Phys., 00, 0000 (1979).

(52) L. Salem, "The Molecular Orbital Theory of Conjugated Systems", Benjamin, 1966.

(53) G. Herzberg and E. Teller, Z. Physik. Chem. B21, 410 (1933).

(54) J. Michl and E.W. Thulstrup, Spectrosc. Lett. 10, 401 (1977).

(55) J. Michl, E.W. Thulstrup, and J.H. Eggers, J. Phys. Chem. 74, 3878 (1970).

(56) E.W. Thulstrup and J. Michl, J. Mol. Structure, in print; J. Michl and E.W. Thulstrup, J. Chem. Phys., submitted for publication.

(57) B.E. Read, Plastics and Rubber; Materials and Applications, Sept. 1976, 123.

(58) G. Aviv, L. Margulies, J. Sagiv, A. Yogev and Y. Mazur, in "Linear Dichroism Spectroscopy", Proceedings of a Nobel Workshop, University of Lund, 1977.

(59) N. Gō, J. Chem. Phys. 43, 1275 (1965).

(60) H.-G. Kuball, T. Karstens, and A. Schönhofer, Chem. Phys. 12, 1 (1976).

(61) E.W. Thulstrup and J.H. Eggers, Chem. Phys. Letters $\underline{1}$,
      690 (1968).

(62) E.W. Thulstrup, J. Spanget-Larsen, and R. Gleiter, Mol. Phys.
      $\underline{37}$, 1381 (1979).

(63) A.C. Albrecht, J. Am. Chem. Soc. $\underline{82}$, 3813 (1960);
      A.C. Albrecht, J. Mol. Spectrosc. $\underline{6}$, 84 (1961).

(64) E.W. Thulstrup and J. Michl, J. Am. Chem. Soc. $\underline{98}$, 4533 (1976).

(65) C.R. Desper, J. Appl. Polym. Science $\underline{13}$, 169 (1969).

(66) A. Saupe, Mol. Cryst. $\underline{1}$, 527 (1966).

(67) R.D.B. Fraser, J. Chem. Phys. $\underline{21}$, 1511 (1953); $\underline{24}$, 89 (1956).

(68) M. Beer, Proc. Roy. Soc. (London) $\underline{A236}$, 136 (1956).

(69) J.H. Nobbs, D.I. Bower, I.M. Ward, and D. Patterson,
      Polymer $\underline{15}$, 287 (1974).

(70) A. Yogev, L. Margulies, and Y. Mazur, Chem. Phys. Letters $\underline{8}$,
      157 (1971).

(71) T.A. Moore and P.-S. Song, J. Mol. Spectrosc. $\underline{52}$, 209 (1974).

(72) Q. Chae, P.-S. Song, J.E. Johansen, and S. Liaaen-Jensen,
      J. Am. Chem. Soc. $\underline{99}$, 5609 (1977).

(73) A. Yogev, L. Margulies, and Y. Mazur, J. Am. Chem. Soc. $\underline{92}$,
      6059 (1970).

(74) T.A. Moore, M.L. Harter, and P.-S. Song, J. Mol. Spectrosc. $\underline{40}$,
      144 (1971).

(75) R.T. Ingwall, C. Gilon, and M. Goodman, J. Am. Chem. Soc. $\underline{97}$,
      4356 (1975).

(76) Q. Chae and P.-S. Song, J. Am. Chem. Soc. $\underline{97}$, 4176 (1975).

(77) J. Sagiv, A. Yogev, and Y. Mazur, J. Am. Chem. Soc. $\underline{99}$, 6861
      (1977).

(78) A. Yogev, L. Margulies, and Y. Mazur, J. Am. Chem. Soc. $\underline{93}$,
      249 (1971); A. Yogev, J. Riboid, J. Marero, and Y. Mazur,
      J. Am. Chem. Soc. $\underline{91}$, 4559 (1969).

(79) A. Yogev, J. Sagiv, and Y. Mazur, Chem. Phys. Letters <u>30</u>, 215 (1975).

(80) A. Yogev, J. Sagiv, and Y. Mazur, J.C.S. Chem. Comm. 943, (1973).

(81) A. Yogev, L. Margulies, D. Amar, and Y. Mazur, J. Am. Chem. Soc. <u>91</u>, 4559 (1969).

(82) L. Margulies and A. Yogev, Chem. Phys. <u>27</u>, 89 (1978).

(83) D. Boyd and P. Phillips, private communication.

(84) B. Nordén, Chemica Scripta <u>1</u>, 145 (1971).

(85) A. Yogev, L. Margulies, B. Strasberger, and Y. Mazur, J. Phys. Chem. <u>78</u>, 1400 (1974).

(86) M. Jordan and C.J. Eckhardt, Lecture at 32. Symposium on Molecular Spectroscopy, Columbus, Ohio, 1977 .

(87) Y. Tanizaki, Bull. Chem. Soc. Japan <u>32</u>, 75 (1959).

(88) Y. Tanizaki, T. Kobayashi, and N. Ando, Bull. Chem. Soc. Japan <u>32</u>,1362 (1959).

(89) Y. Tanizaki and S.-I. Kubodera, J. Mol. Spectrosc. <u>24</u>, 1 (1967).

(90) H. Hiratsuka, Y. Tanizaki, and T. Hoshi, Spectrochim. Acta <u>28A</u>, 2375 (1972).

(91) The relation between T and R_s was first proposed by Smirnov (Zh. Eksp. Teor. Fiz. <u>23</u>, 68 (1952)); he later criticized it (Opt. Spektrosc. <u>3</u>, 123 (1957)).

(92) The understanding of the model is made very difficult by some misprints. In (89) the expression for (III,12) is wrong and in (90), where this error is corrected, the two last equations of expression (1) for $R_d(r)$ and $T(R_s)$ are incorrect.

(93) H. Inoue, T. Hoshi, T. Masamoto, J. Shiraishi, and Y. Tanizaki, Ber. Bunsenges. Physik. Chem. <u>75</u>, 442 (1971).

(94) T. Hoshi, H. Inoue, J. Yoshino, T. Masamoto, and Y. Tanizaki, Z. Physik. Chem., Neue Folge, <u>81</u>, 23 (1972).

(95) T. Hoshi and Y. Tanizaki, Z. Physik. Chem., Neue Folge 71, 230 (1970).

(96) H. Inoue, T. Nakamura, and T. Igarashi, Bull. Chem. Soc. Japan 44, 1469 (1971).

(97) T. Hoshi, H. Inoue, J. Shiraishi, and Y. Tanizaki, Bull. Chem. Soc. Japan 44, 1743 (1971).

(98) H. Inoue and Y. Tanizaki, Z. Physik. Chem., Neue Folge 73, 48 (1970).

(99) T. Yoshinaga, H. Hiratsuka, and Y. Tanizaki, Bull. Chem. Soc. Japan 51, 996 (1978).

(100) H. Inoue, T. Hoshi, J. Yoshino, and Y. Tanizaki, Bull. Chem. Soc. Japan 45, 1018 (1972).

(101) T. Hoshi, J. Yoshino, H. Inoue, T. Masamoto, and Y. Tanizaki, Ber. Bunsenges. Physik. Chem. 75, 891 (1971).

(102) T. Yoshinaga, H. Hiratsuka, and Y. Tanizaki, Bull. Chem. Soc. Japan 50, 3096 (1977).

(103) H. Hiratsuka, Y. Tanizaki, and T. Hoshi, Bull. Chem. Soc. Japan 50, 1282 (1977).

(104) Y. Tanizaki and H. Hiratsuka, Spectrochim. Acta 28A, 2367 (1972).

(105) Y. Tanizaki, H. Inoue, T. Hoshi, and J. Shiraishi, Z. Physik. Chem., Neue Folge, 74, 45 (1971).

(106) Y. Tanizaki, M. Kobayashi, and T. Hoshi, Spectrochim. Acta 28A, 2351 (1972).

(107) A. Fucaloro and L.S. Forster, J. Am. Chem. Soc. 93, 6443 (1971); Spectrochim. Acta 30A, 883 (1974).

(108) B. Nordén, manuscript to be published; L. Gårding and B. Nordén, Chem. Phys. 41, 431 (1979).

(109) K.R. Popov, Opt. Spectrosc. 38, 102 (1975).

(110) K.R. Popov, Opt. Spectrosc. 25, 471 (1968).

(111) N.V. Platonova, K.R. Popov, and L.V. Smirnov, Opt. Spectrosc. 26, 197 (1969).

(112) K.R. Popov and L.V. Smirnov, Opt. Spectrosc. 30, 93 (1971).

(113) N.V. Platonova, K.R. Popov, I.I. Shamolina, and L.V. Smirnov, Opt. Spectrosc. 29, 254 (1970).

(114) K.R. Popov and L.V. Smirnov, Opt. Spectrosc. 28, 610 (1970).

(115) K.R. Popov, Opt. Spectrosc. 35, 607 (1973).

(116) K.R. Popov, Opt. Spectrosc. 36, 65 (1974).

(117) K.R. Popov, Opt. Spectrosc. 39, 142 (1975).

(118) K.R. Popov, Opt. Spectrosc. 39, 368 (1975).

(119) K.R. Popov, Opt. Spectrosc. 39, 285 (1975).

(120) M. Lamotte, J. Chim. Phys. 71, 803 (1975).

(121) C.C. Bott and T. Kurucsev, Chem. Phys. Letters 55, 585 (1978).

(122) N.S. Gangakhedkar, A.V. Namjoshi, P.S. Tamhane, and N.K. Chaudhuri, J. Chem. Phys. 60, 2584 (1974).

(123) R.N. Nurmukhametov and V.A. Lipasova, Opt. Spectrosc. 39, 365 (1975); V.A. Lipasova, R.N. Nurmukhametov, and G.T. Khachaturova, Opt. Spectrosc. 42, 570 (1977); V.A. Lipasova and R.N. Nurmukha-metov, Opt. Spectrosc. 40, 229 (1976).

(124) R. Gleiter, M. Kobayashi, J. Spanget-Larsen, J.P. Ferraris, A.N. Bloch, K. Bechgaard, and D.O. Cowan, Ber. Bunsenges. Physik. Chemie 79, 1218 (1975); J. Spanget-Larsen, R. Gleiter, and S. Hünig, Chem. Phys. Lett. 37, 29 (1976); P. Bischof, R. Gleiter, K. Hafner, M. Kobayashi, and J. Spanget-Larsen, Ber. Bunsenges. Physik. Chemie 80, 532 (1976); J. Spanget-Larsen, R. Gleiter, M. Kobayashi, E.M. Engler, P. Shu, and D.O. Cowan, J. Am. Chem. Soc. 99, 2855 (1977); R. Bartetzko and R. Gleiter, Angew. Chem. 90, 481 (1978). J. Spanget-Larsen and R. Gleiter, Helv. Chim. Acta 61, 2999 (1978).

(125) Y. Tanizaki, T. Yoshinaga, and H. Hiratsuka, Spectrochim. Acta 34A, 205 (1978).

(126) E.W. Thulstrup, M. Nepraš, V. Dvořák, and J. Michl, J. Mol. Spectrosc. 59, 265 (1976).

(127) P.G. Lykos and R.G. Parr, J. Chem. Phys. 24, 1166; 25, 1301 (1956).

(128) J. Linderberg and Y. Öhrn, "Propagators in Quantum Chemistry", Academic Press, 1973.

(129) E.A. Taft and H.R. Philipp, Phys. Rev. 138, A197 (1965).

(130) R.G. Parr, "Quantum Theory of Molecular Electronic Structure", W.A. Benjamin, 1964.

(131) R.G. Parr, J. Chem. Phys. 20, 1499 (1952).

(132) R. Pariser and R.G. Parr, J. Chem. Phys. 21, 466 (1953); J.A. Pople, Trans. Faraday Soc. 49, 1375 (1953).

(133) J.W. Downing, J. Michl, P. Jørgensen, and E.W. Thulstrup, Theoret. Chim. Acta 32, 203 (1974).

(134) P. Jørgensen and J. Linderberg, Int. J. Quant. Chem. 4, 587 (1970).

(135) J. Koutecký, J. Paldus, and R. Zahradník, J. Chem. Phys. 36, 3129 (1962).

(136) R.A. Harris, J. Chem. Phys. 50, 3947 (1969).

(137) P. Jørgensen, Ann. Rev. Phys. Chem. 26, 359 (1975).

(138) A.E. Hansen, Theoret. Chim. Acta (Berl.) 16, 217 (1970).

(139) J. Linderberg, Chem. Phys. Letters 1, 39 (1967).

(140) P.J. Stephens, J. Chem. Phys. 52, 3489 (1970).

(141) L. Seamans and J. Linderberg, Mol. Phys. 24, 1393 (1972).

(142) S.T. Epstein, J. Chem. Phys. 58, 1592 (1973).

(143) E. Dalgaard, Chem. Phys. Letters 47, 279 (1977).

(144) J.A. Pople, Proc. Phys. Soc. (London) A68, 81 (1955).

(145) R. Pariser, J. Chem. Phys. 24, 250 (1956).

(146) J.H. Obbink and A.M.F. Hezemans, Chem. Phys. Letters 50, 133 (1977). A. Kaito and M. Hatano, J.Am.Chem.Soc. 100, 4037 (1978).

(147) N.H. Jørgensen, P.B. Pedersen, E.W. Thulstrup, and J. Michl, Int. J. Quant. Chem. S12, 419 (1978).

(148) H.M. Gladney, Theoret. Chim. Acta 1, 245 (1963).

(149) R.P. Steiner and J. Michl, J. Am. Chem. Soc. 100, 6861 (1978).

(150) I. Fischer-Hjalmars, J. Chem. Phys. 42, 1962 (1965).

(151) J. Michl, J. Am. Chem. Soc. 100, 6801 (1978) and following papers.

(152) J. Michl, Chem. Phys. Letters 39, 386 (1976);Int. J. Quantum Chem. S10, 107 (1976).

(153) A.E. Hansen and E. Nørby Svendsen, Mol. Phys. 28, 1061 (1974).

(154) J. Michl and E.W. Thulstrup, Tetrahedron 32, 205 (1976).

(155) E.W. Thulstrup and J. Michl in "Linear Dichroism Spectroscopy", Proceedings of a Nobel Workshop, University of Lund, 1977.

(156) E.W. Thulstrup and J. Michl, J. Mol. Spectrosc. 61, 203 (1976).

(157) E.W. Thulstrup and J. Michl, Spectroscopy Letters 10, 435 (1977).

(158) J. Michl and E.W. Thulstrup in "Linear Dichroism Spectroscopy", Proceedings of a Nobel Workshop, University of Lund, 1977.

(159) E.W. Thulstrup, Int. J. Quantum Chem. 12, S1, 325 (1977).

(160) E.W. Thulstrup, J. Mol. Structure 47, 359 (1978).

(161) B. Nordén, R. Håkansson, P.B. Pedersen, and E.W. Thulstrup, Chem. Phys. 33, 355 (1978).

(162) T. Dahlgren, Å. Davidsson, J. Glans, S. Gronowitz, B. Nordén, P.B. Pedersen, and E.W. Thulstrup, Chem. Phys. 40, 397 (1979).

(163) N.H.F. Beebe, E.W. Thulstrup, and A. Andersen, J. Chem. Phys. 64, 2080 (1976).

(164) N.H.F. Beebe and J. Linderberg, Int. J. Quantum Chem. 12, 683 (1977).

(165) J.B. Birks, "Photophysics of Aromatic Molecules", Wiley-Interscience 1970, p. 75.

(166) H.H. Jaffe and M. Orchin, "Theory and Applications of Ultra-violet Spectroscopy", Wiley, New York,1962.

(167) E. Clar, A. Mullen, and Ü. Sanigök, Tetrahedron 25, 5639 (1969).

(168) E. Heilbronner, J. Michl, J.-P. Weber, and R. Zahradnik, Theoret. Chim. Acta 6, 141 (1966).

(169) J. Michl, Theoret. Chim. Acta 15, 315 (1969).

(170) J. Michl, J. Mol. Spectrosc. 30, 66 (1969).

(171) G.P. Dalgaard and J. Michl, J. Am. Chem. Soc. 100, 6887 (1978).

(172) M.M. Mestechkin, L.S. Gutyrya and V.N. Poltavets, Opt. Spectrosc. 30, 547 (1971).

(173) I.B. Berlman, J. Phys. Chem. 74, 3085 (1970).

(174) G. Milazzo, Spectrochim. Acta 2, 245 (1944); Gazz. Chim. Ital. 78, 835 (1948); 83, 392 (1953).

(175) For a discussion see G. Herzberg, "Electronic Spectra and Electronic Structure of Polyatomic Molecules", D. van Nostrand, 1966.

(176) H.M. Rietveld, E.N. Maslen, and C.J.B. Clews, Acta Crystallogr. 26, 693 (1970).

(177) J.L. Bandour, Y. Delugeard, and H. Cailleau, Acta Crystallogr. B32, 150 (1976).

(178) S. Ramdas and J.M. Thomas, J.chem.Soc., Faraday II, 1251 (1976).

(179) N.A. Ahmed and A.I. Kitaigorodsky, Acta Crystallogr. B28, 739 (1972).

(180) H. Suzuki, Bull.chem.Soc.Japan 33, 109 (1960).

(181) R.J.W. Le Fevre, A. Sundaram, and K.M.S. Sundaram, J.Chem.Soc. 3180 (1963).

(182) S.P. Gupta and B. Krishna, J.Am.Chem.Soc. 94, 57 (1972).

(183) J. Dale, Acta Chem. Scand. 11, 650 (1957).

(184)   H. Suzuki, "Electronic Absorption Spectra and Geometry of
         Organic Molecules" Academic Press (1967).

(185)   T. Kitagawi, J. Mol. Spectrosc. <u>26</u>, 1 (1968).

(186)   W.G. Dauben, J.I. Seeman, P.H. Wendschuh, G. Barth, E. Bunnen-
         berg, and C. Djerassi, J. Org. Chem. <u>37</u>, 1209 (1972).

E. Rånby, J. F. Rabek

ESR Spectroscopy in Polymer Research

1977. 356 figures, 29 tables. XIV, 410 pages
(Polymers/Properties and Applications, Volume 1)
ISBN 3-540-08151-8

Contents: Generation of Free Radicals. – Principles of ESR Spectroscopy. – Experimental Instrumentation of Electron Spin Resonance. – ESR Study of Polymerization Processes. – ESR Study of Degradation Processes in Polymers. – ESR Study of Polymers in Reactive Gases. – ESR Studies of the Oxidation of Polymers. – ESR Studies of Molecular Fracture in Polymers. – ESR Studies of Graft Copolymerization. – ESR Studies of Cross-linking. – Application of Stable Free Radicals in Polymer Research. – ESR Spectroscopy of Stable Polymer Radicals and their Low Molecular Analogues. – ESR Study of Ion-Exchange Resins.

"... This book is a remarkable example for the successful combination of simplicity and clarity in its tutorial parts and of depth and width whenever and wherever its presents the state of the art... As ultimate and very gratifying reward for his investment the reader gets no less than 2519 references to the literature in excellent alphabetical order.

Springer-Verlag
Berlin
Heidelberg
New York

Scientists who already work with ESR will be greatly assisted in their efforts by this book; those who do not yet use this method will have an easy time to learn and use it. All of them will be grateful to the authors for this exceptional addition to our scientific literature."

J. Polymer Science

Reactivity and Structure

Concepts in Organic Chemistry

Editors: K. Hafner, J.-M. Lehn, C. W. Rees,
P. v. Ragué Schleyer, B. M. Trost,
R. Zahradnik

This series will not only deal with problems
of the reactivity and structure of organic
compounds but also consider synthetical-
preparative aspects.
Suggestions as to topics will always be
welcome.

Volume 1: J. Tsuji
Organic Synthesis
by Means of Transition Metal Complexes
A Systematic Approach
1975. 4 tables. IX, 199 pages
ISBN 3-540-07227-6

Volume 2: K. Fukui
**Theory of Orientation and
Stereoselection**
1975. 72 figures, 2 tables. VII, 134 pages
ISBN 3-540-07426-0

Volume 3: H. Kwart, K. King
**d-Orbitals in the Chemistry of
Silicon, Phosphorus and Sulfur**
1977. 4 figures, 10 tables. VIII, 220 pages
ISBN 3-540-07953-X

Volume 4: W. P. Weber, G. W. Gokel
**Phase Transfer Catalysis in
Organic Synthesis**
1977. 100 tables. XV, 280 pages
ISBN 3-540-08377-4

Volume 5: N. D. Epiotis
Theory of Organic Reactions
1978. 69 figures, 47 tables. XIV, 290 pages
ISBN 3-540-08551-3

Volume 6: M. L. Bender, M. Komiyama
Cyclodextrin Chemistry
1978. 14 figures, 37 tables. X, 96 pages
ISBN 3-540-08577-7

Volume 7: D. I. Davies, M. J. Parrott
**Free Radicals in Organic
Synthesis**
1978. 1 figure. XII, 169 pages
ISBN 3-540-08723-0

Volume 8: C. Birr
**Aspects of the Merrifield Peptide
Synthesis**
1978. 62 figures, 6 tables. VIII, 102 pages
ISBN 3-540-08872-5

Volume 9: J. R. Blackborow, D. Young
**Metal Vapour Synthesis in
Organometallic Chemistry**
1979. 36 figures, 32 tables. XIII, 202 pages
ISBN 3-540-09330-3

Volume 10: J. Tsuji
**Organic Synthesis with Palladium
Compounds**
1980. XII, 207 pages
ISBN 3-540-09767-8

Volume 11:
**New Syntheses with Carbon
Monoxide**
Editor: J. Falbe
1980. 118 figures, 127 tables.
Approx. 490 pages
ISBN 3-540-09674-4

Volume 12: J. Fabian, H. Hartmann
**Light Absorption of Organic
Colorants**
Theoretical Treatment and Empirical Rules
1980. 76 figures, 68 tables. Approx. 260 pages
ISBN 3-540-09914-X

**Springer-Verlag
Berlin Heidelberg New York**

Lecture Notes in Chemistry